# *Moving Up to Medical Sales*

## *From Buyer to Seller*

By Michael A. Carroll

Reedy Press
PO Box 5131
St. Louis, MO 63139
USA



Library of Congress Control Number: 2008920925

ISBN: 978-1-933370-32-3

For information on all Reedy Press publications visit our website at www.reedypress.com.

Printed in the United States of America
08 09 10 11 12    5 4 3 2 1

Artwork by Kevin Belford

"Give a man a fish and feed him for a day. Teach a man to fish and feed him for a lifetime."
—Chinese Proberb

"Find a way to sell boats, fishing rods, and tackle to all these fishermen, and you can feed your family until the kids go off to college."
—Michael Carroll

# *Dedication*

*This book is dedicated to my father, Albert Carroll, who began his medical career caring for injured sailors and Marines and worked his way into the operating room. My dad then made the transition from clinician to buyer, when he took a job in St. Louis as the purchasing agent of a large metropolitan hospital. My father created the concept of integrating purchasing, central supply, receiving, etc., under one managerial umbrella, which he called "Materials Management." My father was truly the first to be formally called director of Materials Management, and he wrote articles and lectured on this new concept early in his career.*

*After twenty years, my dad made the successful transition from buyer to seller when he opened the Kansas City branch of his company. My dad and one other gentleman were the first people to open an entirely new branch to a new market in the Kansas City area. My father was technically the branch manager, but before he had a branch to manage, he had to build the business by meeting potential customers, addressing their needs, and selling them some medical supplies, and he did it as well as it could be done. My father passed away on November 28, 2005, after a long battle with cancer. Part of my success is due to the impact my father had on his customers. When I made my way into medical sales, the mention of my father's name was enough to open many doors that were locked to my competitors at times. To this day, I still run into my father's former customers, and they love to tell me how much they enjoyed working with him and how much they miss him.*

*My father was a very special man in a lot of ways, and much of what I have become today is due to the love of my Mother and Father. Without them, this book would not be possible. A portion of the proceeds from this book are donated to the Lymphoma and Leukemia Society.*

# Contents

# Preface

## *Who Am I to Write This Book?*

Who am I to write this book? That's a fair question, and I hope that by the time you finish the book, you will understand exactly why I felt qualified and compelled to write a guide, or manual, like this. However, if you are now asking this question, I sure hope your boss already paid for the book, but if you decide to throw it in the trash, my publisher still got paid, and therefore, I got paid. Although . . . they told me it might take several months to work off my advance, and I don't really know if returned books count against my sales, but anyway, I am already off the subject.

This book will teach you all the basics of medical sales, but first I want you to know that I was born to sell—plain and simple. When I was born, my relatives tell me that I sold my mom's obstetrician on not spanking me. I convinced him that it was really in the best interest of all parties involved that I should cry on my own, as opposed to being forced to cry due to the barbaric smack of his open hand. Next, I convinced the doctor that if he laid me in my mother's arms, I would stop crying, and voilà, it was done. In this case, it all worked out for everyone in the room. That is the way *I* remember it anyway.

Seriously, I believe I am well qualified to write this

book for many reasons. First, I started selling as a child. I was successful in all of my selling careers, and I sold and managed in the health care marketplace. In addition, as you will learn later, I realized that I could have used a tool like this many times during my medical sales management career.

I truly did start selling at a young age, and here is my story: When I was eight years old, my parents subscribed to *Life, Time,* etc., and they would read the magazines the same week they arrived. Most of the magazines my parents read arrived each week in our mailbox. I did not pay much attention to the magazines myself, except for the occasional issue of *National Geographic,* but I knew my parents seemed to read and enjoy each and every issue.

One day, while irritating my mother at the grocery store, I noticed that the same issue of *Life Magazine* we had in the kitchen at home was for sale at the checkout for seventy-five cents. While we were paying for our food, I asked the checkout clerk if a lot of people bought *Life Magazine.* The startled clerk told me that the store sold out of *Life Magazine* every week. Now, we did not grow up poor—we were exactly dead center of the middle-class—but my parents always tried to teach me the meaning of money. When I wanted something, there were three ways to get it. First, I could wait for my birthday, when I would get one big present. Second, I could wait for Christmas (which was less than a month after my birthday), when my parents were always very generous. And finally, I could earn my own money and buy it. The allowance for an eight year old did not go very far, so that meant I had to go out and get a job. You know, set the alarm, ride the bus to the plant, put in my eight hours, and get a paycheck at the end of the week.

But wait, I was only eight. What kind of a job could I get? Child labor laws were already in place, so my possibilities were limited. I was too small to shovel snow. None of the neighbors wanted to hire someone my age to cut their grass, and I was not allowed to touch the mower anyway. So what was left? I could sell something!

We had a stack of magazines in the living room, and I knew that people liked to read magazines. I also knew that my parents were done reading them, so they were no longer any use at our house. This was it! I could sell the magazines to the neighbors and make a nice little profit: my cost of the magazines = $0.00; selling price, aaaahhhh, = 25¢, 15¢. I did not know what price used magazines would fetch, but I figured if they sold out at the grocery store, they must have *some* value. I asked my mom if it was okay to try to sell the magazines, and she said, "Michael, no one is going to want to buy used magazines, because the news is old in those magazines." I insisted on trying, and she let me. During one balmy Saturday afternoon, I sold out of all the magazines we had stored in our house. Neighbors had paid an average of 15 cents per magazine, and I sold thirty magazines for a total profit of $4.50 (Cost . . . $0.00 . . . Profit . . . $4.50 . . . Priceless). This was a lot of money to an eight year old in 1969. My mom and dad were both surprised. As a matter of fact, I had so much success that now I waited every week for the latest issues to arrive at our house. The first thing I did when I got home from school each day was check the mailbox for any new magazines. If there was a magazine there, I grabbed it, and took it door to door to find a buyer. The magazines fresh out of the mailbox were easier to sell, for even twenty-five cents each.

I happened to hear my mother calling the magazine

company and the post office one day in an attempt to ascertain why my parents were no longer receiving ANY of their magazines. Oops. I forgot about that middleman part concerning letting my parents actually read the magazine before I sold it. I fessed-up to what was happening, but I was just an innocent eight year old trying to make a buck. Can you blame me? No matter how you slice it . . . a salesman was born, or made.

From the used magazine business, I graduated to the lemonade stand, but instead of selling lemonade, I sold Coke *with* ice. Coke was much easier to sell than lemonade. I did not have to make the Coke, I just had to pour it. At five cents per cup, I was able to rake in three to five dollars on a Saturday, and then I could return the Coke bottles and get another five cents for the deposit. I know what you are thinking: This plan has a slight flaw, also. Yes, you are correct. My parents had purchased the Coke, the cups, and made the bottle deposit. This problem, however, was quite fixable. My mother took me to the grocery store with my profits and made me buy the cups and the Coke. This exercise actually helped me understand cost, margin, mark-up, and profit. Keep in mind, I was only nine by this time. As the years went by, I continued to look for ways to sell and make money. (I learned that garage sales are great, but once again, you need to sell stuff that your parents DON'T need.)

At the age of thirteen, my uncle introduced me to MLM, multi-level marketing. This was real selling, at least for the product I had. I actually had to learn a presentation, understand buyer needs, wants, and, mostly, fears (so we could play on those fears). I also had to learn the importance of probing, buying signs, and closing the sale. Believe it or not, this really taught me a great deal about selling as a skill. By age eighteen, I had progressed, or regressed, into the world of retail sales part-time during college. At my store, we actually *sold*. In today's world of retail, I rarely encounter a true professional salesperson, and if I do, I buy from them. My career in retail sales went very well, and I even sold out of many items the store did not think would move.

I continued to learn, study, and sell, and just after my twenty-fourth birthday, I landed my first professional "outside" sales position selling industrial communication equipment—zzzzzzzzzzzzzz. This really was not as boring as it sounds, and I learned a lot, but the best thing about this job was that I got to attend two of the General Electric ROA (Results Oriented Approach) sales training seminars. This is where I acquired additional sales skills and learned to expand upon true "consultative" selling. The experience and skills I learned through my first professional selling job allowed me to enter the world of medical sales at age twenty-six. Now, twenty-plus years later, I am still in medical sales. I am currently a sales vice president covering a large chunk of the country, and I have held positions as high as National Sales Manager and National Marketing Director. I do not mean to toot my own horn, but I would put my listening, presentation, and selling skills up against anyone in this industry, and I feel confident I would prevail. As good as my sales skills are, my experience working with sales people of varying degrees of ability (and some really should never have been allowed to refer to themselves as salespeople), has also been invaluable to improving my own sales skill.

My immediate family helped me obtain the knowledge needed to write this book as well. My wife is a medical sales rep who was first a clinician. My parents both came from the "clinical side" before jumping into sales. My father was a medical corpsman, surgical assistant, and finally the creator of the director of Materials Management position now seen in almost every hospital. My mother started out as a medical corpsman, and then she became a medical and dental assistant, and even a teacher who taught medical assistants before moving into medical sales. Anyway, all these experiences and circumstances have, hopefully, contributed greatly to the information I am going to share with you here. Life is a learning process, and I have never spent one day in medical sales thinking that I did not need to learn anything new, and I would encourage you to do the same.

Enough about me, let's get you started learning how to succeed in medical sales.

# Acknowledgments

Writing a book is not an easy thing, especially if you work sixty to seventy hours a week on a regular basis. Without the help of some very special people, this book would not have been possible. I thank everyone around me for their patience, guidance, support, and understanding.

Thank you to my mom and dad who raised me in an environment where health care was discussed every day. Thank you to my wife and daughter for their love and encouragement. Thank you to Josh and Matt for their guidance; and thank you to the following people who helped make this information available to you: Ivan Rothman, Doug Brown, Kurt Krieghbaum, Chris Simpson, Bill Bascom, Andy Mills, Dr. Carlton Turner, Dr. Andrew Scaduto, Tom Huling, Alex Liberman, Mike Lee, Cynthia Fleck, and all the reps who read and offered opinions on portions of this book.

# Chapter One

## *Introduction and Background (The First Mystery)*

Medical sales is a continuously growing, expanding, and evolving field that requires a continuous influx of qualified sales people. Unfortunately, few monoliths of higher learning offer anything remotely resembling a guide, not only in how to sell, but specifically, how to sell in the medical arena. Some "experts" would have you believe that "selling is selling," regardless of the product, service, or forum. There is some truth to this belief, but anyone who has spent time selling medical supplies or medical services (from this point forward lumped together as "medical sales") will shake their head at the thought of medical sales being anything similar to selling to other institutions. First of all, the medical field is not, contrary to popular belief, dominated by physicians. The same physicians who spend all the years in college, medical school, internships, residency, specialty residency, and fellowships are usually not the people driving the bus when it comes to determining which medical supplies and services a hospital, nursing home, or home care agency will utilize. Sorry to shock you, but this is true.

The opposite is true for pharmaceuticals. With a pharmaceutical drug, generally, a physician order is required, and the docs have the vast majority of the decision-making power when it comes to which drug is used on which patient. This book does not help guide you through the skills of pharmaceutical sales, because there are none. Ouch! I say that with my tongue grazing the inside of my mouth, but there is some truth to what I say. I do not want to dwell on this, but in today's world, the pharmaceutical rep goes through an incredibly grueling training process to learn everything anyone could ever know about one or two simple conditions and one or two drugs. When they finally get out on their own carrying a bag, they are limited to mostly providing lunch in the doctor's office for everyone and dropping off samples with a signature. If you are just entering pharmaceutical sales and someone gave you this book by mistake, see if you can pass it on to someone else. You may, however, find some of this information useful for your next job, when you graduate to a position in medical sales. (Yes, that is a dig at pharmaceutical sales.)

When it comes to selling medical supplies, especially in the acute care setting (hospitals), most of the decisions on what to buy, what to use, what to stock, and on whom to use the medical supply, are made by nurses, clinical nurse specialists, advanced practice nurses, physician assistants, and therapists. These are the people on the front lines of patient care who are usually in close contact with the patient and their condition. Do not misunderstand, doctors are key to all health care settings, but when it comes to medical supplies, most physicians are quite comfortable leaving the decisions for what brace, butt cream, cover dressing, tape, vital system monitor, scale, glucose meter, specialty bed, feeding pump, glove, X-ray machine, IV start kit, drape, pre-packaged bathing system, and gauze pad a hospital wants to use in the hands of the senior nurses and therapists. (More on this later.) The one exception may be in the surgical arena. Hospital purchasing, contract managers, OR nurses, and surgical techs do have some decision-making power, but the surgeons are fairly particular about the products they use on and in the patient while in the operating room. I want the surgeon to be my friend!

When it comes to medical supplies, most of the aforementioned products are either Class I FDA exempt products, or they are some type of 510K medical device. These products do not generally go through an IRB (Investigative Review Board)-type study process (with the exception of implantable devices), and the FDA does not regulate their sale in the same way. This may help to explain why the physician's input is not quite as great as it is with OR products and pharmaceutical drugs.

The selling of complete intangibles in the form of medical services is a huge health care market. A medical service is just like the name says—a service, as opposed to a supply or device. The sale of medical services is key for many areas of the health care marketplace. In the United States, more than 5,000 home health care agencies are licensed to provide care to patients leaving the hospital. How do you think the hospital or physician or discharge planner recommends where that patient goes? The sales rep for the home care company *sold* the facility or doctor on their quality of *service*. Nursing homes do deliver a product, in a way, but their product is the service/experience/care of their nursing home. Therapy companies sell their services, as do home medical equipment companies, imaging companies, diagnostic testing companies, etc. So, knowing how to sell a service to the health care market is as important as knowing how to sell a product, or to whom to sell this product or service.

The bottom line here is that you will be selling a professional product or service to, primarily, health care professionals. Whether you are calling on a nurse, physical therapist, physician, dietitian, or pharmacist, you will be talking to someone who has a good understanding of the area your product covers, and they expect the information you deliver to be up to date, accurate, and reliable. These people also understand medicine, the human body, patient care, and products. Long before you enter the health care facility in an attempt to sell something, you have to become the professional salesperson your customer expects. I am sure you have all heard the adage, "Know your customer, know your product, and know your competition." This is especially true when it comes to selling medical supplies. If you are trying to convince the wrong type of caregiver to use your product, they may not fully understand the appropriateness of your product, which could result in your product being used on the wrong type of patient. If the caregiver does not clearly understand how your product is to be properly applied, they may use it in a way that could harm the patient or perhaps make the condition worse. Furthermore, if the health care professional is accustomed to using a competitive product, and your product differs in some way, they could use the product inappropriately, which could also harm the patient. For these reasons, and so many others, you absolutely, positively have to know your customer, your product, and the competition inside and out.

A side benefit to knowing all about your customer, product, and competition is that you are much more likely to sell more products. The more you know, the more you can help your customer, and the more of an asset you are to them. The better prepared you are for questions, the more successful you can be at convincing that caregiver to give your product a try, and the better your presentation will be. Finally, your knowledge and the ability to use it will build your credibility in the eyes of the health care professional.

If you are already a health care professional and you have decided to move to the other side of the desk, you may be a little bored by certain parts of this book, like the description of the layout of the acute care facility, but maybe even *you* will not find the politics of each department interesting. I do believe, however, that this information can be a helpful review for you, and it may trigger some memories and inquiries that will be of some help later on. Try to bear with me and the other readers in order to lay the groundwork for what is to come as you begin *your* career selling in the market where you once practiced.

*****

I wrote this book because I found a need in my business for this type of information. As a regional manager, I have to hire qualified sales people. In the division of the company for which I work, we have a semiformal hiring hierarchy when searching for a candidate, and this is also my preference. The first level is hiring someone from one of our direct competitors. This makes for a very easy transition. The new rep just has to learn the new products, while everything else stays the same. The second level is to hire someone with experience in our narrow market sometime in the recent past. The third choice is to hire someone with a medical sales background, but not selling our type of products in our market. The final level of candidate is either a person with a strong sales background but with no medical sales experience, or a clinical person (nurse or therapist) with no sales experience. The other managers in my company and I have all hired people from each level who have been successful, and we have also hired people from all levels who have not lasted.

In my interviewing experience, I have come across many candidates who had all the skills I wanted in a rep but no medical sales experience, and they would not know a Med/Surg floor from the giftshop. I have also hired clinical people who had absolutely no sales experience. The hardest part of hiring either one of these types of candidates is the time that it takes to teach them what is in this book. Trust me, it is a long, long process, especially if they come from sales but not medical sales. Hopefully, you can now see why I wrote this book. I hope that this book will give the managers in the medical sales arena a tool to bring new people up to speed. Medical sales professionals have to spend a great deal of time reading and learning about their product and the competition. I do not cover every single area in great detail, because I want you, the reader, to understand this market without boring you. I trust that you—the future medical sales success story—will find the information contained herein helpful, and I wish you all the best of luck!

# Chapter Two

## *Layout of the Health Care Facility (The Second Mystery)*

You may ask, at this point, why the guts of this book start with such a dull subject as Layout. Read on, and you will understand. Before you make your first sale, you have to know where to go, right? When I say, "layout," I am referring to the physical layout of a health care facility in addition to who does what. Since the hospital reflects the most intricate layout, I will cover it first. Then we will deal with nursing homes, home care, hospice, etc. For those of you who are nurses, therapists, or some other clinician, you are quite familiar with the layout of a hospital, so this will just be a little refresher. If you have a clinical background, you will undoubtedly say, "My hospital has the blah, blah, blah on the second floor, not the ground floor." This section is meant to be a general guide, and there WILL be exceptions. Remember, I am the one with the book contract, so just read!

At the start of my career in medical sales, I had the northern half of Kansas and the northwestern one fourth of Missouri. A year or so later, I changed to the eastern half of Missouri and the southern half of Illinois. I knew the St. Louis area well, since I grew up there, but I did not really have much familiarity with Southern Illinois or the rest of the state of Missouri. The great thing about calling on hospitals is that if you can find the town, you can find the hospital—all because of those wonderful, little blue signs with an "H" on them. I do not know of another industry that tells you where your primary customer is located in every town across America, but if the town has a hospital, you can usually find it. One caveat, however—in some small towns, teenagers like to play with the arrows from time to time, but by and large, the signs will take you where you want to go.

## *The Hospital*

Let's start with where to park. Hospitals are getting more and more crowded, and there are more and more cars on the road. Therefore, more patients plus more cars equals fewer parking spots at the hospital. Some hospitals will have a spot designated for sales personnel parking or service people. If they do not, which is usually the case, try to park as close to the Purchasing, or Materials Management Department as possible. If you do not know where the Purchasing Office is located, you are not doing a very good job with knowing your customer. In most hospitals, sales people have to check in with the Purchasing Office before seeing anyone in the facility. Most Purchasing/Materials offices are located on the bottom floor close to the receiving dock in the vast majority of today's hospitals. You will learn more about the Materials Department in the next section.

Let's take a look at the general, floor-by-floor, layout of a typical acute care facility starting from the bottom up.

H

## Ground Floor

*Materials/Purchasing*
*Central Supply/Central Sterile*
*Food Services/Dietary*
*Housekeeping/Waste Handling*
*Receiving*
*Storeroom*
*Laundry*
*Engineering and BioMed*

Why would one suppose that all of these areas are traditionally located on the ground floor? The answer is obvious. Products are received on the dock, checked by Purchasing/Receiving, sent to the storeroom, and upon request, shipped over to central supply awaiting stock requests from the floors. It just makes sense to put all these integrated areas near each other and on the same floor, which happens to be the ground floor. Many hospitals now locate the Materials Office on a different floor and even in an entirely different building sometimes blocks or miles away from the actual hospital. But generally, Purchasing is located on the ground floor.

What about Housekeeping and Dietary? Why are they on the ground floor? Once again, it really has to do with the location of the street. In many facilities, Dietary receives its own product through a separate receiving area, so the ground floor makes sense. Since Dietary is usually on the ground floor, you will many times see the food preparation area and even the cafeteria on the ground floor. In today's modern hospitals, however, the cafeteria tends to be located on the lobby level or main floor to make the cafeteria more accessible to patient families and visitors who pay full price for the food.

Housekeeping and Waste Disposal operate under a similar formula, but their entire product is going out instead of coming in. Housekeeping is responsible for cleaning all areas of the hospital, including spills of bodily fluids. Hopefully, bodily fluid spills are handled as medical waste, and housekeeping has to bag them in separate red bags indicating "Bio-Medical Waste." Normal garbage collected by housekeeping is disposed of in a dumpster, which of course, is located on the street level. Medical waste is also picked up by a garbage-type truck, hence the need for street access. Carrying bags labeled "Bio-Medical Waste" or "Bio-Hazard" through the front doors of a hospital could cause some concern on the faces of patients and visitors walking in. Just an FYI, bio-medical waste has to be processed in a special way, so it cannot be lumped in with the everyday trash. Hospitals pay a much higher fee for processing bio-medical waste, so it has to be (supposed to be) kept separate.

Engineering and Bio-Medical Engineering (Bio-Med) have always been in related areas, mostly due to the role of engineering in the hospital. Bio-Med uses bio-medi-

cal technicians, and Engineering uses engineers from a variety of disciplines, including facilities, mechanical, electrical, and others. Well-managed facilities keep the engineers and technical people together so they have someone else to bore—I mean to talk to . . . a little geek humor there. Be that as it may, engineering is responsible for keeping a hospital lit, heated, and air conditioned, and all of those systems are accessible on the ground floor, kind of like your basement.

In the old days, visitors to a hospital for diagnostic tests would have to go to the ground floor for X-rays and other imaging services. According to the National Council on Radiation Protection and Measurements[1], the fear of radiation contamination made it necessary to use special walls, floors, and ceilings for the X-ray equipment. Locating X-ray equipment on the ground floor alleviated the need to insulate the floor. With modern medical technology and construction, including lead-lined gypsum board, Radiology services no longer are banished to the basement. You will usually find Radiology on the first, second, or third floor in a hospital, or maybe even in a free-standing building near the hospital.

Lastly, you will probably find the Laundry in the basement. Just as the laundry area in American houses has shown a shift from the basement to the main floor and even the second floor of our homes, Laundry departments in today's modern hospitals have altered tradition as well. Laundry has not necessarily moved to different floors; instead, many hospitals now either utilize an off-site laundry facility or they outsource their laundry services.

What do you say we get our heads out of the basement and climb the steps to the next floor?

## *Lobby Level or Main Floor*
### *(Usually -#1, L or M on your elevator panel)*

| | |
|---|---|
| *Front Desk* | *Lobby* |
| *Information* | *Patient Accounts* |
| *Human Resources* | *Cafeteria* |
| *Gift Shop* | *Emergency Room** |

**Sometimes on the Ground Floor*

I assume that the reader can make the connection here between these areas and their location on the first, or main, floor, but I still want to "elaborate in order to educate." Patients and visitors come in through the main door, they enter the lobby, and they usually first go to the front desk or information desk. If people are visiting patients, they will often visit the gift shop and the cafeteria. When patients leave and have to return to discuss their expensive bill, Patient Accounts is the place to go. Finally, if someone from the outside wants to apply for a job at the hospital, they will enter the main door, go to the information desk, and then to Human Resources, so it just makes sense to locate these areas in close proximity to the entrance and on the first floor.

When people are sick or injured to the point that the injury or illness necessitates a trip to the Emergency Room (ER), hospitals try to make it easy, once again, on the patients and their loved ones. The ER is commonly located in an easily accessible place, which allows for cars and the ambulances to drive up close to the hospital. The Emergency Room, although sometimes located on the ground floor, is usually found right off the street, at "street level" with signs clearly marking its home.

The Emergency Room is also the area where you will find the ER docs, nurses, and staff. The ER can be a difficult place to see clinicians, but it really depends on the day or night you are there. The ER gets medical supplies from Central Supply and their drugs from the hospital Pharmacy, like other areas of the hospital.

## *Second through Fifth, Twelfth, or Fifteenth Floor*

*Medical/Surgical (From flu to post-gall bladder surgery)*
*Recovery*
*Intensive Care Unit (ICU)*
*ICU/CCU Step Down Unit*
*Pediatrics/Nursery*
*Education (Usually an office)*
*Wound Care/Enterostomal Therapy (Usually an office)*
*Surgery*
*Physical Therapy*
*Occupational Therapy*
*Cardiac Intensive Care Unit (CCU)*
*Telemetry*
*Oncology*
*Infection Control (Usually an office)*
*Radiology*

These floors are all commonly referred to as "the floors," and there is no general rhyme or reason to their location within a hospital. When a hospital is built, administrators, with the help of engineers and architects, will determine which unit or specialty area goes where. The floors are interchangeable, so if the Oncology floor started off as a small area and the hospital expanded its cancer treatment areas, the decision could be made to move Oncology to a different area or floor. Maybe a hospital opened twenty years ago with limited cardiac surgeries and then hired a popular cardiac surgeon, and his team now performs fifty open-heart surgeries per month. The administration for that hospital would probably dedicate a separate area just for post-op care of the cardiac patient. These are just two examples why the hospital directory signs in the hallways can come in handy.

If you are going to be truly successful in medical sales, and you sell to hospitals, you have to get out on the floors and meet with the folks that actually use your products or products like yours. Up on the floors is where you will find all the sick and injured people, the people recovering from surgery, and people having diagnostic tests performed. Every hospital is different depending on their area of emphasis, but generally speaking, the floors where you find people who are sick, having tests, or recovering from general surgeries will be called Med / Surg floors, or Medical / Surgical floors. Although these floors are quite commonly full of medical and surgical patients, specialty areas can sometimes be found here.

Surgery, Recovery, and Radiology, are usually *not* considered "the floors," but in fact, they are usually on a floor above the first floor, and they need to be discussed along with everyplace else. The following specialty areas of "the floors" are explained along with a rating of the ease of selling and selling opportunity on a scale of one to ten. One is not good (harder to sell to), and ten is really, really good (easier to sell to). This rating system is just one person's opinion (mine), but I have consulted with many of my friends in the business about it, and they generally agree. The following are the areas and the rating I give them:

## *Specialty Areas*

| | *Ease of Selling* | *Selling Opportunity* |
|---|---|---|
| *Surgery* | *5* | *10* |

When I say *Surgery* I mean the Operating Room: The place where the doctor actually cuts open the patient and removes or fixes whatever is the order for the day. Even though some hospitals have specialty surgery areas—Cardiac Surgery, for example—most surgeries, regardless of type, are in the same general area. The well-appointed hospital will have multiple surgery "suites," or rooms, which just means that more than one surgery can occur simultaneously. Believe it or not, I have been in hospitals with areas to accommodate seventy-five patients. With the exception of a hospital with a large cardiac surgery practice, the OR staff could be working on any type of surgery at any time. They may have an appendectomy at 7:30 and then scrub in on a 10:45 tumor removal. We will talk more about the specialties of surgery when we discuss politics.

All surgical-related products and services are used and conducted out of the OR area. Most people that work in the OR never leave the area, except to eat, and even then, if the food is not brought in by a company or rep, sometimes they do not get time to eat. OR time is very expensive, so everyone tries their best to keep on schedule—if that means the staff does not have time to go to the cafeteria, they don't go. In surgery, you will also find . . . ta-da . . . the surgeons. For many of you, these are people you will really need to meet, befriend, and understand. Most surgery areas have a lounge where staff can rest, read journals, listen to us reps, and even eat.

This is a great place to conduct business with a surgeon or surgery staff. If you find a surgeon who will not meet in the OR area, you will have to catch him or her at their private office, where your time will really be limited.

| | *Ease of Selling* | *Selling Opportunity* |
|---|---|---|
| *Recovery* | *4* | *1* |

The recovery area is always located close to surgery for a couple of reasons. First, the patients leave the OR to wake up from a general anesthetic, and they need to get to that secure, and relatively clean, area as quickly as possible. Many things can happen to patients in recovery while they come down off the anesthesia, so close monitoring is important. In addition, occasionally the patient has surgical complications in recovery. If this happens, patients have to be rushed back to the OR, and the closer an operating room is to recovery, the quicker this can be accomplished. To be quite frank, very few selling opportunities exist in the Recovery Room, unless you have a product specifically designed for Recovery, and even then, the docs and OR people will probably be involved in the decision-making part of the process.

## *Radiology*

| | *Ease of Selling* | *Selling Opportunity* |
|---|---|---|
| *Diagnostic* | *6* | *7* |
| *Therapeutic* | *7* | *7* |

In my section on Politics, I will discuss this further, but medicine has two different types of radiology: diagnostic and therapeutic. Diagnostic radiologists are physicians who are specifically trained to analyze complex pictures created through X-ray, MRI, CT scans, PET scans, cardiac catheterizations, etc. Therapeutic radiologists are involved in using the power of radiation to treat conditions like cancer. In most larger hospitals, the MRI machine, CT scanner, and X-ray machines used by the diagnostic people (these are commonly referred to as diagnostic imaging machines) are grouped in the same area, and similar to the surgery area, there is usually a place where the radiologists can read films and congregate. In most of the larger hospitals we see today, the cardiac catheterizations (this is when they pop a hole in your body, usually in the leg, and they snake a little probe up a vein all the way to your heart to check for blockage) usually occur separately in the Cath Lab. Unlike in the OR, radiologists do not have to stay in Radiology all the time, but they often do, because many have so much film to read. Some will come into Radiology only when they have a patient's films that need to be seen. It just depends on the hospital and the radiologists' practice. For every actual radiologist at the hospital, there will be two to three radiology technicians, or techs. The techs actually do most of the work performing the imaging procedure, and then the actual radiologist reads the results. Catching a particular radiologist in the actual Radiology Department can be tricky, due to the odd schedules they sometimes keep. If you are looking to sell a $2 million scanning machine, I am sure you will find them. The majority of therapeutic radiology takes place in a separate, free-standing building near the hospital or in an entirely different location.

| | *Ease of Selling* | *Selling Opportunity* |
|---|---|---|
| *Intensive Care (ICU, CCU, SICU, and NICU)* | *4* | *8* |

The Intensive Care Unit (ICU) and Cardiac Care Unit (CCU) are usually located on the same floor, and it is common for a patient to go from the CCU to the ICU before going to a Step-Down Unit. By the way, when some clinicians talk about a patient being "up on the unit," they are usually referring to one of the ICU areas. The nice thing about the ICU or CCU is that they tend to have their own waiting rooms. This is a great place to go and quietly kill time between in-services (more on this later), but you will almost never get ICU or CCU staff to meet with you in those waiting areas. The only difference between the ICU and CCU is the type of patient, as you probably already figured out. The CCU has the cardiac patients that need intensive care. Intensive care means just that—care that is intensive. The ratio of nurses to patients in an ICU can be as low as one to one, whereas on the normal Med/Surg floor that ratio is commonly ten or fifteen to one. Some hospitals have a post-surgery intensive care area (SICU).

An Intensive Care Unit, cardiac or otherwise, allows for extremely easy patient monitoring. A typical ICU is laid out so that the nurse's desk runs the length of the unit, and this desk is usually a busy place. From the nurse's station, a physician or nurse can see most, if not all, of the patient rooms, and the patient rooms usually are all glass, which allows the staff the ability to see directly into the room. Intensive care requires not just constant electronic monitoring but also visual monitoring of the patient. These patients are very sick, and the nurses need to be able to monitor the slightest change in the patient's condition. When you have to do in-servicing for ICU or CCU staff, you will usually have to do it right at the nurse's station so that you do not pull the staff away from the monitoring area.

The NICU, which you will hear called the "nick-you," can be a tough area to see from an emotional perspective. The NICU treats premature babies, or babies that are born with one of several different conditions that require close monitoring. The NICU is usually in close proximity to the Nursery and/or Pediatrics, and this unit has an entirely different monitoring system from the ICU. Since these patients are babies, they do not take up the kind of space that an adult ICU patient needs, so several babies in warmers or incubators can fit into one large room, which looks similar to a nursery. The layout of the NICU lends itself well to accommodating many nurses, doctors, and our little patients.

*Telemetry/Step Down* *See ICU*

If a hospital has a Telemetry Unit, it is often near, or a part of, the ICU or CCU area. Telemetry is the practice of monitoring patients on a constant basis. The patients in Telemetry may be from the OR, or ICU, and they are hooked up to a bunch of monitors. Sometimes Telemetry will be part of the ICU, and other times Telemetry will be an area unto itself. At times, a patient on a Med/Surg floor will develop some kind of complication that necessitates the monitoring of a critical vital sign. That patient would move to Telemetry, or they could stay in their room and move under the monitoring of the Telemetry technicians. Some of the most sophisticated hospitals in the United States have Telemetry departments that can monitor every patient in that given hospital. The Step-Down Unit is an area where patients can receive monitoring and stepped-up care, but it is a level below the intensity of the ICU.

| | *Ease of Selling* | *Selling Opportunity* |
|---|---|---|
| *Nursery and Pediatrics* | *4* | *6* |

If the NICU is perhaps the saddest area of the hospital, the Nursery is one of the happiest. As everyone knows, the Nursery is where babies in good health go after they are born. The Nursery, in today's hospital, is often found within a special area called Obstetrics, or even a "Women's" or "Women's and Children Center." The clinical folks caring for babies are in and around the Nursery, along with all the happy and gloating parents, grandparents, aunts, uncles, etc. If a hospital has a Nursery, they will also have a Labor and Delivery area, and—you guessed it—this is where babies are delivered. Labor and Delivery is always located close to the Nursery for obvious reasons.

Not all hospitals have a specific Pediatrics Unit, but most teaching facilities will have a Pediatrics area. You should know what we mean by pediatrics, but just in case, pediatrics treats children. Pediatrics will usually cover patients from infants to age eighteen, although some hospitals have different ages when they pass the pediatric patient to an adult Med/Surg area. Pediatric units are usually bright and cheery, with the warmth that make children feel safe and loved. Do not be surprised to see nurses and even the doctors wearing bright, colorful, animal-print, or cartoon-print scrubs and smocks.

| | *Ease of Selling* | *Selling Opportunity* |
|---|---|---|
| *Oncology* | *3* | *8* |

In smaller facilities, Oncology patients will be located within the general population on Med/Surg floors, but in larger hospitals Oncology patients will be on specific floors. There is good reason for this. When patients are undergoing cancer treatment, many times their immune system will become so weak that they lose the ability to fight off any type of infection. Special precautions have to be taken with these patients, and it makes it easier to group them together on the same floor. Also, the drugs used to fight many of the cancers and unique medical issues created from cancer treatment are likely to be used in the same area, which makes it better for the patients and more convenient for the staff. In some facilities, you will also find patients with diseases of the immune system lumped in with the oncology patients. These specialty patients include those with auto-immune diseases like HIV.

Unlike the actual Oncology floor, some hospitals will have a specific area called "Radiation/Oncology." This is actually a *treatment* area, as opposed to a patient-care area. Oncology refers to the total treatment of the cancer patient. Radiation therapy involves the use of highly concentrated beams of radiation to destroy cancerous masses inside the body, like I described in the section on Radiology. Some patients only have radiation therapy, and some patients only have chemotherapy (the use of drugs to fight cancer), but many cancer patients undergo both types of treatment, and you can find them for treatment in the Radiation Therapy or Radiation/Oncology area of the hospital. In today's health care market, commonly a Radiation Therapy Center or even a "Cancer Center" will be a free-standing building away from the hospital. Some centers are not even directly affiliated with a hospital. Due to the nature of using radiation for treatment, the Radiation Therapy area has to be very well insulated to contain the actual radiation. Specific radiology docs work in the Radiation Therapy areas, and you will also find people known as radiation therapists (RT). The RT is a technician, much like in the X-ray and imaging area. The techs do have some say-so, and overall, the radiation therapists are very approachable. Do not mistake the Radiation Therapy docs with the radiology docs. The two do not practice the same kind of medicine.

| | *Ease of Selling* | *Selling Opportunity* |
|---|---|---|
| *Education (Usually an office)* | 7 | 2 |

If a hospital does not have a specific Education Department inside the facility—all teaching facilities will—then they will probably have a person who carries the responsibility of coordinating staff and patient education. These people are very easy to see, and their office could be located on any floor of the hospital or even at an off-site location. One duty, which will be discussed later, is the job of educating staff when the hospital adopts your new product. This type of education will usually take place in the area where your product will be used, and this is called an in-service. Basically, the "in-service" occurs when the facility brings your product IN SERVICE at the hospital. You may also have to in-service the staff that will be using your products when you start an evaluation for your products. The Education Department will be key, along with your champion, in coordinating the schedule for whomever you will have to educate and when. In the eyes of the Education Department, this in-service is paramount to the staff understanding how they are to use your product. In reality, this is your big opportunity to increase the sales of your products in that facility. We will talk more about this later, but in-servicing is really a sales tool rather than an education tool. Regardless, the Education Department will be a key point of contact.

| | *Ease of Selling* | *Selling Opportunity* |
|---|---|---|
| *Infection Control (Usually an office)* | 7 | 7 |

Most hospitals have an Infection Control Department (IC), which may include nurses as well as physicians and PhDs. The Infection Control Office can be located anywhere in the hospital on practically any floor. A hospital that is part of the "100,000 Lives Saved" or "Surgical Care Improvement Plan" initiative will have an active and involved Infection Control Department, and they are important people to get to know for many types of products and services.

The Infection Control Department will usually have an Infectious Disease doctor—either an MD or PhD—who is ultimately responsible for the department. Despite this fact, the IC nurses are the ones who really run the show. This type of nursing specialty requires a higher degree of training, and in today's market with nosocomial (hospital-acquired) infections everywhere, the profile of Infection Control has seen a parallel increase in their visibility.

For any products that remotely have anything to do with the control or prevention of infections, you want to involve the IC nurses. Believe it or not, a relatively few number of companies sell products aimed at the control of infections and bacteria. Those companies who do sell to the IC nurses are able to establish strong relationships with these clinicians, and you do not want them on the other side. At least one representative from Infection Control will sit on the key hospital, patient-care committees, and these types of committees can often determine whether or not your product becomes a part of that hospital's protocol. These nurses can be very nice, but they can also be quite political, partially because they can pull a lot of weight. Be nice to them, and be their friend.

| | *Ease of Selling* | *Selling Opportunity* |
|---|---|---|
| *Wound Care/Enterostomal Therapy (Usually an office)* | *6* | *8* |

The Wound Care Office or Enterostomal Therapy Office tends to be a one- or two-person office that deals with problems related to chronic wounds, ostomy, and incontinence. The nurses who staff these offices used to be called ET nurses prior to 1992, but they are now CWOCNs, which stands for Certified Wound Ostomy and Continence Nurse. In addition to caring for wounds or skin problems, these nurses care for patients who have had to undergo surgery to replace the normal flow of the bowel or bladder with a plastic bag worn outside of the body. These well-intentioned nurses will work with patients who have wounds on their body while in-patient, and many times the CWOCNs will see the same, or additional, patients on an outpatient basis. When it comes to a patient with a stoma, the patients will usually be seen initially as in-patients following their surgery, and then the CWOCN will follow up with them as an outpatient when they have problems.

| | *Ease of Selling* | *Selling Opportunity* |
|---|---|---|
| *Physical Therapy* | *6* | *8* |

Physical Therapy (PT), which can also be called Rehabilitation Services or "Rehab," is just like it sounds. The PT Department is staffed by certified physical therapists and therapy techs who help patients with the physical recovery after an injury, illness, or surgery. Physical therapists go through an education and training program that takes longer to complete than a nursing program. The program for therapy technicians is shorter than the nursing program. The Physical Therapy, or Rehab, area tends to be located on the lower floors, but not always. Partly due to high patient traffic, the therapy areas are a bit more accessible than other parts of the hospital.

| | *Ease of Selling* | *Selling Opportunity* |
|---|---|---|
| *Occupational Therapy* | *6* | *8* |

Much of the information in the preceding paragraph is also true for the Occupational Therapy Department (OT). The main difference between these two departments is what they are trying to accomplish. In a very broad sense, occupational therapists, or OTs, work with patients to rehabilitate their fine motor skills, such as writing, eating, picking up objects, while physical therapists concentrate more on gross motor skills and larger muscles and muscle groups important for things like walking.

| | *Ease of Selling* | *Selling Opportunity* |
|---|---|---|
| *Laundry:* | *5* | *5* |

I must confess that I have never spent much time working with the Laundry Department or the people working with third-party laundry operations. When rating these areas for ease of selling and selling opportunities, I heard a wide range of opinions. Overall, reps I talked with seem to feel that the Laundry Department in the hospital was very approachable, but the department has most of their decisions made in Materials Management. I am told that a free-standing laundry operation is a great selling opportunity and the people are very approachable, although many of the companies in this business are still mom-and-pop type operations, especially in smaller metro areas.

# Chapter Three

## *Politics*

Now that you understand the layout of a hospital, and hopefully you are starting to get some ideas about where your products may fit within the layout, you now need to understand the true goofiness of the politics in health care. Every health care facility is a little different, and some of you who have a clinical background may once again say, "My hospital wasn't like that." But just be open-minded and try to understand that the politics discussed here should be considered the general and common observation from my point of view and the view of other reps and managers in the field. Let us not forget who has the book contract!

From a purchasing standpoint, the buyer is the most important thing for you to understand. Hospitals, and many nursing homes and home care agencies, have both a clinical and a fiscal buyer. In a nursing home or home care agency, the clinical buyer is usually the director of nursing (DON), the asistant director of nursing (ADON), or a clinical specialist like a wound nurse or treatment nurse. Some larger homes and agencies may have their own wound care nurse or infection control nurse. The fiscal buyer, which may be a purchasing agent or an office person, is usually given the task of ordering the products, but they usually do not have a great deal of decision-making power. Hospitals, on the other hand, are a totally different animal. Hospitals have many clinical buyers and, oftentimes, many fiscal buyers.

The purchasing task in a hospital is almost always centralized to one department: The aforementioned Purchasing Department or Materials Management. The buyers in the Purchasing Department have the sole job of ordering products and services for the hospital. Larger hospitals will have multiple buyers who split their responsibilities by product lines or companies. I have personally visited hospitals with ten different buyers in one hospital. When my father started in the business, he was the sole purchasing agent, and he reported directly to the administrator of that hospital. By the time my dad left the acute care setting, three people carried the purchasing function in his hospital, and he had the final say-so on any difficult issues.

Maybe it will help if I illustrate what happens from the time a product is boxed at the manufacturer until the box is delivered to the hospital. This example illustrates a typical transaction cycle for selling a product into the hospital:

> Let's say Mary Martin is the clinical nurse manager for the ICU at St. James Hospital. Mary tries to keep her department running in tip-top form, and she tries to continually provide the highest quality care for her patients. Mary attends the AACN meeting, which is a national meeting attended by nurses who practice in the critical

care arena. Mary attends a lecture by Dr. Jonesberg who has developed, tested, and studied a new type of infusion pump. Mary sees the potential advantage to using this new type of pump, so after the lecture, Mary visits the exhibit hall to find the company who makes this new pump. The Acme Corporation has a booth at the convention, and Mary looks for the pump in the booth. There it is! As Mary approaches the pump, a representative from Apex approaches Mary. Mary asks, "Is this the pump discussed during Dr. Jonesberg's lecture?" The nice representative tells Mary that this *is* the pump and this is all the other stuff you need to know about the pump. After a long discussion with the well-dressed gentleman from Apex, Mary *has* to try one of those pumps in her ICU.

When Mary returns to her job at St. James on Monday morning, she calls Bob in Purchasing and tells him that she has to get the new Apex Infusion Pump in her ICU. Bob asks her if she has used one yet and how much it costs. Mary, unfortunately, has not used one, and she has no idea how much it costs, but she has the name and phone number of the local Apex rep, Dave, which she gives to Bob. Bob calls the Apex rep and asks for more information. Like any good rep, Dave tells Bob that pricing is dependant on several factors, but the product is available. Bob asks the rep if the hospital can evaluate the new pump in Mary's ICU, and Dave says, absolutely! Bob, Mary, and Dave set up a time to bring in the pump so that St. James can evaluate the performance of this new product.

Dave from Apex brings this new infusion pump to Mary and her staff for the evaluation. In order to fairly evaluate the product, Dave must educate all of the nurses who will be involved with using the product for the evaluation period. During the one-week evaluation, everyone learns how to use the pump and gets to try it out. Mary loves the pump, the staff loves the pump, and even the doctors do not mind the new pump. So now Mary wants one, and she wants it bad. As a matter of fact, she has to have this new pump, and ICU care, as they know it, cannot continue without this new pump. Guess what? Mary, the clinical buyer, is sold! Done deal! The pump is ordered, right? Wrong.

Mary tells Bob that the evaluation went great, and the department would like to have the new pump as soon as possible. Actually, they will need twenty-two new pumps to replace all their existing Draconian pumps. Bob knows that Mary wants the pump, and Dave from Apex knows Mary really wants the pump, but Bob wants to get the pump at the best possible price, so Bob calls Dave. Many things can happen at this point, but for the sake of our story, Dave and Bob meet. Dave tells Bob that each pump is $1,000. Bob wants to know if the pump is on their GPO agreement (more on this later), and what the pump costs on that agreement. The pump is so new that it is not on any national buying agreements. Bob asks Dave about volume pricing. "If we order a quantity of the pumps, can we get a better price?"

Dave explains to Bob that they *do* have volume discounts and that if they order twenty-five to fifty pumps, they can get a price of $900 each. Bob only needs twenty-two pumps, so he tries to get the twenty-two at the twenty-five price. Dave, being the good sales rep he is, tells Bob that he would love to help him, but the pricing is quite firm, due to the overwhelming demand.

At this point, Bob can try to beat up Dave to get a discount from the $1,000 level, or he can decide to up the order to twenty-five pumps to get the better price. After some tense negotiation and Dave's many "calls to his manager and his manager's manager," Dave says that the best Apex can do is $950 per pump if they buy only twenty-two. Bob is faced with two options:

A. Bob can buy 22 pumps @$950/pump
Total Cost $20,900
Or
B. Bob can buy 25 pumps @$900/pump
Total Cost $22,500

Essentially, if Bob buys the twenty-five pumps, he is getting the extra three at a cost of $533 each. Another way to look at it is that he is getting one and a half pumps for free if he buys the twenty-five. In this case, Bob knows that pumps go down, and patient census goes up and down, and so he makes the decision that it is worth the extra $1,600 initial expenditure to get the extra three pumps, but he negotiates sixty-day terms to pay for the pumps.

Truthfully, this scenario could have worked out several different ways, but the bottom line is that Bob did the fiscally sound thing as the fiscal buyer to get Mary her pumps and get the best deal he could for the hospital. Little does Bob know, but Dave and Apex would have gone to $925 per pump for twenty-two if the higher price would have kept the sale from going through. Little does Dave know, however, that Bob would have paid list price if he had to. To review: Mary's job as the clinical buyer was to get the best product for her patients and staff, from a clinical point of view. Bob's job was to get the product Mary felt she needed at the best possible price and at the best possible term for the hospital. Dave's job was obvious—sell as many pumps at the highest profit margin possible.

As the reader can see, the clinical buyer can be a nurse manager, specialist, a physician, a therapist, or even the head of housekeeping. But the fiscal buyer is almost always the buyer or purchasing agent, or even the director of Materials Management. With this format, hospitals maintain a check-and-balance system. If it were up to the clinical buyers, they would get everything they want or think they need, and if it was up to the fiscal buyer, they would always buy the least costly alternative. For most health care institutions, this process works pretty well, until it comes to the committee.

In many of today's hospitals, there is an additional step, which involves getting the product approved through a "committee." Hospitals have all sorts of committees: Value Analysis, Skin Care, Surgical Services, Infection Control, etc. The previous scenario may have been a little different if the pumps had to be approved by a committee. We will discuss committee selling later, but committees meet at certain times, most commonly once each month, and if the pump had to go through committee, Mary may have had to present her evidence as to why she needed the pump, and the committee would have to approve the purchase. Sometimes, other people on the committee may say that they want to evaluate the pump, before a decision is made, and this process can slow things down from a week to months.

Two areas that are outside the box for the purposes of this discussion are Surgery and Prescription Drugs. I will not spend a lot of time on drugs, because remember: This book is not for pharmaceutical sales. Almost all new drugs for a hospital have to go through the Pharmacy and Therapeutics Committee (P&T) before the drugs are brought in-house. Once a drug is in-house it can be ordered by anyone who writes a prescription. The Surgery, or OR, Department can also be its own unique entity.

Many hospitals, especially teaching institutions and those with docs that perform a large volume of surgeries, treat the OR and the Surgery Department as an island unto itself. If this is the case in one of your hospitals, the OR will have its own budget and its own purchasing person, usually referred to as the clinical resource manager (CRM). Some hospitals have recently adopted the use of a CRM in the main hospital Purchasing Department, which is a good idea because it gives purchasing a clinical perspective. The OR, however, has utilized this type of position for years. In the OR, products are usually presented to the docs, chief OR nurse, or physician's assistant to evaluate, and then the CRM takes if from there if the clinicians want or need the product.

But wait . . . the box holding the pump is not yet

sitting on the shelf in the hospital? Apex still needs to get the pump delivered to the hospital. Apex may ship the products directly and not use distributors, but this is more the exception than the norm. If Apex sells and ships direct, they will take a purchase order from the hospital and ship the product. The more likely scenario will have Apex selling the product through the hospital's distributor of choice. Apex will get the ordering information to the distributor, or the information may get to the distributor through the Purchasing Department. Either way, Apex ships the product to the distributor, and they may charge the distributor the per-unit $900 price. The distributor will add their markup, or distribution fee, to the original cost of the product. Once the distributor receives the pump, they will likely ship it to the hospital as part of their weekly or daily order. Sometimes the distributor may have Apex "drop-ship" the pump to the hospital. This means that Apex will ship the product to the hospital and send the invoice to the distributor. Either way, the box containing the pumps will get to the hospital so that Mary can finally start using it to provide improved patient care.

*****

Since this chapter is titled Politics, I want to make sure you all understand the politics that dictate behavior and decision-making both clinically and in purchasing. No five hospitals are the same, but what follows are the general politics of different departments based on twenty-plus years working with every department within the modern health care facility.

The bottom line is this: Almost everyone, whether

clinical or non-clinical, has an agenda. The vast majority of health care professionals and purchasing people have the patient, the staff, and the facility at the heart of any decision they make, but what follows are some general guidelines that you may find true in your hospitals. The guidelines are listed by department with the level of politics you will commonly see. What do we mean by politics? In this case, politics refers to all the extraneous things outside of the sales cycle that contribute to what may or may not affect the sale, meaning it may not just be about the product and the price. (1 is non-political, and 10 is very political. A higher or lower rating is absolutely no indication of the quality of patient care.)

*Medical/Surgical (from flu to post–gall bladder surgery)* *1*

Floor nurses, charge nurses, and even nurse managers are traditionally not heavily involved in the decision-making process and tend to be rather apolitical. These nurses, even the nurse managers, are just trying to care for the patients as well as they can under the direction of hospital policies.

*Surgery* *6*

Surgical staff generally do what makes the most sense, but reps benefit greatly from relationships established here. If you are able to penetrate the OR, or if your product is specifically an OR product, then you will need to know that many times your sales calls will be from 5:30 to 6:30 a.m. The surgeons and PAs commonly see reps before their cases start. Another important thing to note is that surgery people tend to hang with surgery people.

*Radiology* 2

Radiologists are traditionally open-minded and have somewhat free reign for getting what they need. If you sell a diagnostic imaging tool, you will want to get to know the techs in Radiology. These people will be key to getting your product in front of the right radiology doctor, whether we are talking about diagnostic or therapeutic radiology.

*Intensive Care (ICU)* 6

Most critical care areas are great about improving patient care with new and better technologies, but some critical care managers have rather large agendas. The nurse managers in the ICU are decision makers, however, and if they want your product, they have the ability, in most hospitals, to ram it through. Sometimes their decision on what product they use can determine whether a patient lives or dies that week or day, and they are fully aware of this.

*Cardiac Intensive Care (CCU)* *See above*

*Neonatal Intensive Care (NICU)* 2

The docs and clinical specialists make most of the decisions in this area, and they want to do whatever they can for their patients, no matter the cost or obstacles. If the NICU wants something, they will get it 95 percent of the time.

*ICU/CCU Step Down Unit* *See Med/Surg*

*Pediatrics/Nursery* 2

Many products for the Pediatric and Nursery market are only available or only intended for babies. If a product falls into this category, the sale can be relatively easy because of a lack of competition of adult-used products. You may find that if your hospital has a NICU, the nurse that heads the NICU also makes decisions for the nursery.

*Oncology* 4

Oncology is rarely a political area, but much of what they use is drug-oriented, and they are used to pharmaceutical company treatment, which means non-stop lunches provided by the reps and many dinner presentations. The problem with this scenario is if you have a non-pharmaceutical product, the doctors and staff still expect to get it for patients outside the hospital like they do the drugs they use. They are used to writing a prescription, giving it to the patient, having the patient fill the prescription at the pharmacy, and getting the drug. Not all over-the-counter products and medical devices are available this way, so you may have to take another route.

*Education (Usually an office)* 2

These wonderful people do not get to wield substantial influence, and they usually work in support of everyone else in the hospital to help educate staff and patients for various issues. Since this office works in support of the patients and the rest of the hospital, they do not usually get involved in product selection, but when your product is evaluated or adopted by the hospital, you will work with them. You need to be nice and supportive of their efforts, because if you do a poor job with education, they will convey this to the clinical decision makers, which never helps.

*Infection Control (Usually an office)* 4–7

This may seem a little odd to rate something as a range as opposed to a number, but Infection Control (IC) is a unique area. Some IC nurses recognize their role as support of proper infection control guidelines for the facility. Other IC nurses feel that they need to get involved in many product decisions that are not truly under their area of authority. Because of the importance of infection reduction in today's hospital, IC nurses can exert influence in certain areas at certain times. Once again, if they will ever be on your radar screen, get to know them and befriend them. They tend to be a fun and knowledgeable group.

*Wound Care/Enterostomal Therapy (Usually an office)* 8

Wow! Talk about politics. We could spend an entire chapter on the politics of wound care and the ET nurses. After surveying numerous sales reps who call on ET nurses, the message is fairly consistent. If you are one of their best friends, you will be able to maximize your penetration with products they use. If you do a great job supporting their ostomy patients, you may get preferential treatment, even when it comes to non-ostomy-related products. On the other hand, if you are new and not working for one of their favored vendors, it will be an uphill battle, and you better hope your products are on contract with these facilities if you want to get any business at all.

The CWOCN/ET nurses are near and dear to my heart, because I have been calling on them since I began my career in medical sales. In most facilities, these nurses can ram a four-foot-wide product through a one-inch hole, so to speak. Of course I am speaking metaphorically, but it is important to remember that they can move mountains when they are motivated to do so. A product that is not appropriate or not as good as something priced competitively can find its way into the hospital through the ET nurse if she really wants it to happen. In addition, a company can lay all the best groundwork during the evaluation process, be priced better, and have better clinical documentation, yet the ET nurse can stop this product, which is supported by so many people, from landing in the facility. The ET nurse is primarily involved with a very narrow scope of products, which are usually limited to those that affect wound care, skin care, ostomy, or incontinence. When they are involved with a particular product, however, they can make you or break you.

Establish relationships with them. Most ET nurses do not like the hard sell, and they believe that they should not use just one manufacturer for their products. I believe the prime reason these nurses can carry so much weight is that everyone else in the hospital is afraid of wound care and ostomy care. Purchasing certainly does not want to get involved with wounds, because wounds are nasty, disgusting, smelly things that should be left to the experts. Consequently, when the ET nurse wants something, she usually gets it, but that is changing, primarily due to the proliferation of Group Purchasing Organizations (GPOs) and the compliance therein.

*Materials/Purchasing* *8*

Even though they tend to be better than they used to, this is a department that definitely plays favorites. I have personally been involved with a situation when our company had a better product, clinically, that was supported by clinical staff. Our product had a better price, was on contract, and was a better option for patient care, and we still could not get the business, because a purchasing person or materials director did not prefer our company. Keep in mind, also, that these people can have memories like an elephant. If your company did something they didn't like twenty years ago, they may never forget it and hold it against you forever. One of the reps I know likes to say, "Did you hear that the Beatles broke up?" Fortunately—or unfortunately, depending on your contract status—a great deal of the purchasing choices have been narrowed due to Group Purchasing Organizations (GPOs). If your products are on contract, the Purchasing Department may have some difficulty ignoring you, but if your products are not on contract, you may have a tough time selling water to a nurse dying of thirst in a hospital where your company does not have strong relationships in the Purchasing Department.

*Food Services/Dietetics* *4*

Many of the Dietetic decisions are made outside the department, but the skilled registered dietitians (RDs) are very non-political. However, the people involved with food service can get a tad political. Only a handful of companies provide food and food service products, so the competition is narrow and very much based on price and service. Dietary managers who usually supervise all the food service needs for the hospital, from the patient meal trays to what is served in the cafeteria, like to be entertained, and once you earn their trust, you will have a friend for life. The registered dietitians, on the other hand, provide nutritional support to the patients, especially those with complex diseases requiring specialty nutritional needs. They may, however, supervise the Dietary Department. Only a few product and service categories exist that require contact with the RD. If you do have to meet with the RDs, please know that they have at least five years of school and had to undergo a grueling internship in order to sit for the RD exam. Many dietitians now use a new designation of RD, LD—registered dietitian, licensed dietitian. Many people, especially in the world of gym rats and personal trainers try to call themselves nutritionists or dietitians without undergoing such a rigorous educational program, so the National Dietetic Association added the LD designation in the early 1990s.[2]

*Housekeeping/Waste Handling* 1

Housekeeping directors are very down-to-earth people in most cases, and if you show them a better mousetrap at a decent price, they will buy it. In addition, they sometimes have their own budget.

*Engineering and BioMed* 2

The Biomedical Department is staffed by engineers and techs, primarily, and they tend to be reminiscent of the MIS Departments of any large company. Engineering, on the other hand, can have a bit of a "construction contractor" mentality, which can, every once in a while, make decisions a bit political, but by and large they are an easy department to approach.

*Therapists (PT/OT)* 2

I am always amazed how accommodating the people in the therapy departments can be. As I explained earlier, the therapists are well-trained and well-educated, and they want what is best for their patients. I have found them to be more than willing to try new products and technologies, and they will give you quick and honest feedback on you, your product, and your company.

*Laundry* 4

Remember, knowledge of the Laundry Department is not my personal strong suit, so I have to rely more on the opinions of other sales reps for this area more so than any other. The one common thing that I have discovered about the Laundry Department is the way they do their contracting. Often, the contracts are signed one year at a time, which may be a tradition, but it may also be reflective of how quickly service can change for a laundry service provider. The people that tend to sell to free-standing laundry contractors are reps who sell textiles. Textiles is a broad term for all the fabric, primarily non-disposable, items at the hospital, such as towels, blankets, sheets, etc. Scrubs and uniforms can go to Laundry as well. Laundries look at a very narrow scope of bullet points when making a decision on their textile provider: Cost of item, quality (thread-count) of the material, and variety. Many times, the hospital will make all the decisions on the towels, blankets, scrubs, and so on, then contract the laundry services to a third party. In a third-party laundry that does not purchase their own textiles, the selling opportunities are very limited.

*Teaching Hospitals* *3–9*

Teaching hospitals are facilities that train medical interns and residents to be full-fledged doctors. Because they are teaching institutions, these large hospitals will try to maximize the exposure of their residents to new and different things, but these same residents are able to write orders for your products, and they will also remember your products when they get out on their own. Some teaching hospitals make it difficult for reps to do their jobs, but once you get something in a teaching facility, it is harder to get it out. You should get to know the residents, because they will be full-fledged doctors soon, and they are there to learn. The residents will learn what they hear from senior staff, but they will also learn what they hear from a sales rep, like you.

Teaching hospitals are often followed, or preceded, by the word *University,* but not always. A teaching hospital is a facility that has, or is associated with, a medical school. Where you have medical schools, you have medical school students. Where you have med students, you have interns. Of course, where you have interns, you have residents, and where you have residents, you have doctors teaching students. What does this mean to you? If you develop loyalty in the residents, they will be more likely to use your products for the rest of their careers. No joke! I come across physicians every week who are using archaic products and combinations of products they learned about when they were residents when people used the horse and buggy for transportation. Even though it drives reps and managers crazy at times, if I can get them hooked on my product while they are residents, maybe some guy will be writing a book forty years from now complaining about doctors who are still using the old "Michael Carroll's product" technology. Hey, things could be worse.

Keep in mind that with teaching facilities, the doctors are truly there to learn, and the more you can do to facilitate that learning experience for the staff, the better chance you have of getting the ear of those young, green doctors. Just a side note: These residents always seem to be hungry. If you feed them, they will listen.

# Chapter Four

## *Anatomy of a Nursing Home*

People tend to have strong opinions about nursing homes and the care provided therein. If you are able to spend a substantial amount of time in a nursing home, you will come to appreciate the difficult job the staff experiences on a day-to-day basis, and you will appreciate the differences from one home to the next. Like most businesses and health care facilities, there are some that are better than others. Some nursing homes I would trust with my mother, and some I would not allow my worst enemy's demented aunt to enter. Today, nursing home care is a bit more consistent due to the tight state and federal oversight and large monetary settlements for poor patient care in recent years, and some nursing homes are as nice as a five-star hotel. Nursing homes vary widely on their layout and amenities, but the vast majority will fit within some of the generalities listed below.

### *Layout*

The physical layout of a nursing home differs widely from home to home, so I will not spend a great deal of time discussing this aspect; however, the layout of the level of care is important to understand. Every nursing home, or long-term care facility as they are commonly called, is a little bit different. Depending on the geographic area, socio-economic area, payer mix, and age of the building, nursing homes can range from country club–type retirement homes to long-term intensive care to substandard minimal care. Long-term care facilities are divided into five basic categories, which are all based on the type of care provided at that particular facility: assisted living, residential care, intermediate care, skilled care, and sub-acute care.

## *Assisted Living*

These facilities provide little or no actual nursing care. Think of these as the true "retirement homes" people used to refer to when they meant a "nursing home." The assisted living facilities (ALF) are generally nice in appearance, provide activities and privacy, and are usually expensive. If an apartment in a normal apartment complex could be rented for $1,000 per month, the equivalent apartment at an ALF complex could be easily in the $3,000–$6,000 range. Granted, some meal service may be available, but the level of nursing care is almost nonexistent. This is not a bad thing; this is what the resident prefers. Most ALFs have some way of contacting a nursing desk if a medical issue comes up.

## *Residential Care*

The residential care facilities, or RCFs, are quite similar to ALFs, but they usually provide the housing and care in one building. The building could look like a traditional nursing home, but the patients have their own small private mini-apartments. The Catholic Church owns many residential care facilities across the country meant to house the elderly nuns upon their retirement. In a typical RCF, the presence of nursing care is a bit more evident than in an ALF, but some people cannot distinguish the two. The reader may find two nearly identical residential care facilities in a given city, and one facility may call themselves an RCF and the other may consider themselves an ALF.

## *Intermediate Care*

As patients age, or when their health starts to decline, they may need care that is a step up from the residential-type care. This care is provided by an ICF, or intermediate care facility. The ICFs provide a higher level of care to address most patient health issues, but they do not provide quite the level of care required in a skilled nursing facility. Intermediate care facilities have started to lose their growth, and they have been replaced by more mixed-patient facilities offering different levels of care in the same building.

## *Skilled Care*

When most people picture a nursing home with lots of sick people or elderly patients with failing health, they are picturing a skilled nursing facility (SNF). Skilled care is just that, care that is skilled, or at least more intensive than an intermediate care facility. Skilled nursing facilities have twenty-four-hour nursing care with a registered nurse in the facility at all times. SNFs can provide a full range of services for some of the frailest patients in our society. Skilled nursing facilities can provide IV infusion care, enteral feeding, catheter care, and many other services. Typically, a skilled nursing facility will have Physical Therapy, Occupational Therapy, and even Speech Therapy visiting on a daily basis, or they may even have their own in-house Rehabilitation Department.

## *Sub-Acute*

The highest level of long-term care outside of a hospital is provided by a newer brand of nursing home called a sub-acute, or long-term acute care, facility. The sub-acute facilities started in the form of hospital-type buildings that specialized in long-term ventilator-assisted patients. These facilities have evolved to the point where they are providing almost all levels of hospital-type services except for surgeries. Vencor, now called Kindred, was one of the early pioneers of this type of care, and other companies like Select Specialty Hospitals, Regency, Promise, and Intensiva have joined the business of providing high-level care. It is interesting to note that Select Specialty and Intensiva most commonly will lease hospital space on a particular floor directly from a hospital to provide their long-term sub-acute care. For instance, St. Joseph Hospital may not be using the east wing on the fourth floor, so Select Specialty will lease those thirty beds from the hospital and provide their own care, from products to staff, for those patients.

*****

You may be asking, why are there so many different levels of care, and why can't they all be in one location? Good question. The answer is multi-fold. First, many facilities do provide all levels of care from the ALF to the sub-acute in the same building. The different levels of care are typically divided by wing or section of the facility. In some of the nicest, largest, and most expensive long-term care facilities, there could be an entire complex that will allow a patient to buy into the ALF section and be guaranteed all levels of care as their health warrants. This type of care continuum is gaining in popularity all over the country. Second, the different levels of care are reimbursed at different rates. Private insurance and Medicare pay less for assisted living than they do for sub-acute, for obvious reasons. An efficiently run long-term care facility can make money on each level of care.

In addition to the levels of care and reimbursement, the payer mix, which is the type of insurance the patient carries, can determine a great deal about the appearance and amenities of a long-term care facility. If a facility has all "private pay" patients, which means the patients have long-term care insurance or their family, themselves, or their estate are paying all the bills; the facility can charge whatever the market will bare, because their rates are not based on government reimbursement for the level of care. A home that has all Medicare Part A beds will only get Part A reimbursement for 100 days. After the Part A ends, the patient would become a private-pay or Medicaid patient. The ability of the home to be fiscally responsible can allow them to make money on the Part A beds to help subsidize the money they do not usually make on the Medicaid patients. Conversely, a home that is Medicaid-only is receiving the lowest reimbursement possible, and they cannot often afford many of the luxuries of a private-pay home. Be that as it may, a smart and prudent nursing home owner can provide a high level of quality care and still make a small profit in a Medicaid-only home, but they have to work at it and run a very tight ship.

## *Politics of the Nursing Home* 3–9

The turnover of key decision makers is higher in a nursing home than just about any area of modern health care. The old story amongst us reps used to be, "If that DON (Director of Nursing) would not buy your product, come back next week, and present it to the new DON." It is not quite that bad, but a bit of turnover in long-term care facilities is not uncommon. As a brief overview, it is important for you to understand the key people in a nursing home and their traditional roles:

### *Administrator*

The administrator of a nursing home is the coach and driving force behind the operation. The administrator is a key decision maker, especially where finance is a consideration. Ultimately, the administrator is the person who gets the credit or blame for anything good or bad that happens at a nursing home. The administrator is usually hired by the owner and will usually hire the DON, who is the next key person down the line.

## *Director of Nursing (DON)*

This is the key clinical decision maker who has the very difficult job of providing quality care. If the administrator is the coach, the DON is certainly the quarterback of the facility. The DON, who must be a registered nurse, has the responsibility of supervising patient care, hiring the staff, working with state and federal regulators—called surveyors—and being on call practically 24/7 for anything and everything that happens at the nursing home. The job of DON is a tough one, and this is why turnover is common.

## *Assistant Director of Nursing (ADON)*

You can probably figure this out, but this is the second clinical person down the food chain, and he or she acts as backup to the DON. The ADON is a key player, and in many cases, a nursing home is only as good as its ADON. The ADON may be an RN, but he or she could also be a licensed practical nurse (LPN). Depending on the individual nursing home, the ADON may have a great deal of decision-making authority.

## *Treatment Nurse*

The treatment nurse often has the assignment of completing individual patient specialized treatments, such as wound care treatment for all or many patients. This person does not, traditionally, have much decision-making power, but he or she is doing a difficult job, and if they do not like something, they will be very vocal to the ADON and the DON, which can lead to a change in product usage. In some cases, you will want to work closely to make sure the treatment nurse is using your products properly, and also make sure he or she knows how to contact you in the event of a product issue.

## *Purchasing*

The majority of nursing homes, believe it or not, do not have a dedicated purchasing agent. In most facilities, the purchasing function is carried out by a person whose primary function is something else. The medical records clerk, receiving clerk, or even the operations manager may have responsibility for ordering and receiving supplies. In most nursing homes, the worker with purchasing and/or receiving duties is not a key decision maker. With that being said, you still want to keep them in the loop on anything important. Remember, they are a customer too, and developing relationships with customers can never hurt. If you are calling on a larger nursing home that has a dedicated purchasing person, they would be no different than the purchaser in the hospital, and the same issues apply.

## *Dietary Manager*

Most nursing homes do not employee a full-time registered dietitian (RD), as we discuss in the next paragraph, but almost every state requires the home to have a certain number of hours of consulting completed by a registered dietitian. The registered dietitian function is, by and large, relegated to reviewing the chart and weight information of their patients, especially the patients who are being fed by an enteral feeding tube. If you need to call on the RD for the facility, you may have a tough time, because they may only spend a few hours each week in the home. Other than recommending certain types of feeding products, the RD is not going to be a decision maker for the home.

The cooking and food preparation function, however, is performed by internal staff. The dietary manager supervises the kitchen, the cooks, and most aspects of food preparation for the residents. The majority of facilities give the food ordering and selection function to the dietary manager who works in tandem, sometimes, with the RD who visits that particular facility.

In a nursing home, most of the specialty functions that are quite common in a hospital are filled by consultants and agencies, providing functions from physical therapy to X-rays to occupational therapy. Most nursing homes cannot afford to employ full-time physical therapists, occupational therapists, dietitians, etc., so they contract out for these services on an as-needed basis. Most therapy services are paid for by the nursing home under their daily rate, but some services are actually billed directly to Medicare or private insurance. Despite the part-time nature of these therapists and consultants, the importance of getting to know them, if they are involved with the use of your product, cannot be over-emphasized.

# Chapter Five

## *Going Home to Home Care*

Layout and politics can be covered in a short space for home health care. Home health care agencies provide care in the patient's own home when they are sick. Therefore, the nursing staff is not providing care in their own building, like a nursing home. Home health care nursing has been around since the pioneer days, long before we had long-term care facilities. If a patient's care needs can be met in their own home, the patient is happier, in most cases, and the cost of care is greatly reduced.

Home care agencies will have an office where the billing and ordering function is performed. In addition, most home care agencies will stock products that the nurses need at one central location. Many home care agencies have weekly, bi-weekly, or at least monthly meetings where all nursing staff are required to attend. These meetings are great opportunities to provide product sales, education, and training for the whole staff at once. Home care agencies normally welcome companies to come in and provide staff education or product training, especially if the representative can provide continuing education credits and/or goodies, like cookies, donuts, snacks, etc. Unfortunately, some of the products that are used by the agency are not left up to the choice of that agency. Many times, the referring physician will order a specific product for his or her patient. Depending on the relationship between the home care agency and the doc, the home care agency may be able to substitute a product they have in stock, but sometimes they are stuck purchasing what the doctor ordered. This becomes especially difficult for the agency when a home care patient goes to a wound center or some other therapy service. The doctor there may order Product A, which the home care agency orders, but then a week later, the doctor may change the order to Product B. In this case, the home care agency may have had to buy a box of the dressings, and now they are stuck with nine of them that will hopefully be used by some other patient down the road. Anytime you can help a home health care agency with a difficult physician, you will score some major brownie points.

The home care agency will usually consist of an administrator who runs the operation, a director of nursing (DON) who directs and oversees all clinical staff, and many times an assistant director of nursing (ADON), who assists the DON. The majority of home care agencies will have at least one office person who is responsible for the billing, answering phones, placing orders, receiving products, scheduling patient visits, and collecting money. A larger agency may have a separate person for each one of these functions, but if it is a small agency, that office person is a key ally.

The way a home care agency makes money will be discussed under reimbursement in the next chapter, but by and large they operate under the rules of reimbursement by Medicare. Occasionally, they are lucky enough to have private-pay patients whose insurance companies tend to pay a bit more than the government. Due to the flat-rate, "episode of care" reimbursement you will learn about in the next chapter, home care agencies must operate in a fiscally sound manner. The quicker they can nurse the patient back to health or fix the patient's health issue, the easier it is for the agency to make money. In the home care arena, the highest cost to the agency is the soft cost of the nursing visit, which averages $50 to $150 per visit. If the average reimbursement is less than $2,500, the cost of visits can eat up the reimbursement at lightning speed. If you can show the home a way they can reduce costs by decreasing the number of nursing visits or getting patients better quicker, you are providing a valuable service to the agency.

One specialized area of home care is called hospice. Hospice care is for a patient with a terminal diagnosis, and it can be home care or in a facility. Typically, the hospice agency is providing care for a patient in the end stages of cancer. My own father had hospice care, and the nurses that provide hospice care are special. Think about it, every patient you, as a nurse, would be caring for is going to die in a relatively short period of time. Nurses, because of their nature, bond easily with their patients, especially in the home setting, and it is difficult to bond with someone who is about to die. Hospice care is provided not to heal or cure a patient, but to provide comfort. This type of care is known as palliative care. If a home care patient has a wound, the home care agency is trying to heal that wound as soon as possible. With hospice care, the nurses are not trying to heal the wound;

they are trying to make it less painful, provide reduction of odor, and decrease the burden on the family.

Keep these differences in mind when calling on hospice care. With all of that being said, the structure of a hospice agency is almost identical to that of a typical home care agency. One new wrinkle in the hospice equation is a new type of care referred to as free-standing, in-patient hospice. A hospice nurse cannot be at a patient's home full time, and if there are no family members available to help meet a patient's needs when the nurse is gone, the patient has to be institutionalized. That may sound bad, but it is not. This is no different than a home care patient who gets too sick to be cared for at home, so they go to a nursing home, and some nursing homes have special in-patient hospice beds where a hospice agency nurse comes to the nursing home and provides that hospice care. The new free-standing hospice is like a nursing home, but all the patients in the facility are under hospice care. These hospice facilities tend to be smaller, partly because of the high patient bed turnover. In a typical nursing home, patients may be admitted for care that could last for years and even decades. With hospice care, the facility knows that in a short time the patient will expire and the bed will be able to be occupied by a new patient.

# Chapter Six

## *Distribution (Who Moved My Box?)*

In health care, almost everything comes in a box. Your job, as a medical sales professional, is to convince your customer to buy whatever is in your box. Once you have done the great job following all the rules of consultative selling and walking your product through the process, your job is done . . . Or is it? How do the contents of your box—in your company's warehouse—get to your customer? In this case, health care is not much different from other industries. Distributors, generally, move the boxes from point A to point B.

This broad discussion on health care distribution is intended to help you understand the process of getting your product directly into the hands of the end user. Manufacturers make the products, and manufacturers usually have sales people to sell the products. If a manufacturer does not have their own sales force, they may contract with an independent manufacturer's representation company, or they may sell 100 percent through a distributor utilizing the distributor's sales force to move their products. Either way, the product has to get to the health care facility somehow. Some manufacturers, especially in capital equipment (stuff that is more expensive, non-disposable, and lasts a long time), sell direct to the facility. There are a couple of manufacturer/distributors that make or contract-manufacturer buy their own products for direct sale to the health care facility. Although, the most common scenario is one of manufacturer—distributor—health care facility—end user.

For the sake of discussion, let's use the most common scenario as our assumption for how the product will get to our nurse Mary's ICU. We already discussed the sales process and cycle. Purchasing placed the order for the product, but from whom did they actually order? For the sake of argument, we are going to assume that Dave's IV pump is sold through common distribution channels. Even though Bob, our purchasing agent, agreed on the price with Dave, Dave is going to take the order through the hospital's preferred distributor, also referred to as their prime vendor. The distributor purchases the product from the manufacturer, has the product shipped to the distributor's warehouse, and then ships that product, along with the rest of St. James' weekly order, to the hospital. So, the Apex box goes from Apex to the distributor. Then that same Apex box is loaded from the distributor's shelf to the distributor's truck for delivery to the hospital. The Apex box arrives at the hospital, along with the rest of the delivery. Someone in the Receiving Department checks in the box as part of their order and places it on the hospital shelf, ready to be delivered to the ICU for Mary to use. Simple! What is in it for the distributor, you may ask? The distributor has a markup for moving the box. This markup is getting lower and lower, but it is usually determined up front with a contractual agreement at 3–7 percent on top of the price of the product.

## *Who Are the Distributors?*

In the 1960s, every city had several medical distributors. Hospitals and nursing homes bought from whomever they liked, or from whomever gave the best price or the best service. With all the changes in payments and profit margins, some distributors flourished more than others. The larger distributors started buying up local distributors and consolidating operations over the next forty-plus years. In the market today, you will find only a handful of national distributors for hospitals, as well as only a handful of national distributors for long-term care. Some smaller, regional distributors have managed to hold off competition from the bigger guys, but with all the GPO agreements, some of the regional distributors who are not approved GPO distributors, have to claw and scratch for what they get, while the national guys just keep getting bigger.

The largest national distributors in today's market for acute care facilities are Cardinal (formerly Allegiance), Owens & Minor, and Medline Industries. On the nursing home side, McKesson, Gulf South, and Medline are the major national players. For pharmacy distribution, the only major national players left are McKesson Drug, Cardinal Drug, and Amerisource-Bergen. You may ask, how did all this consolidation come about? Well, I am going to tell you . . .

## *Cardinal (Acute Care Distribution)*

Cardinal Health has its roots, believe it or not, in the food distribution business. In 1971, Cardinal was a food distributor and then entered the drug distribution business in the early 1980s after purchasing an Ohio pharmacy supplier.[3] Primarily through acquisition, Cardinal grew to be one of the largest drug wholesalers by the early 1990s. The move into Med/Surg distribution happened like this: Baxter bought American Hospital Supply, who was already a major distributor, and then Baxter became Allegiance. Cardinal, who was formerly a major drug wholesaler, bought Allegiance. Allegiance was already a giant player in the acute care distribution market, but with Cardinal's purchase of Allegiance, Cardinal became the largest acute care distributor in the United States and maintains that position today.

## *Owens & Minor*

Owens & Minor claims its roots to be in the early part of the twentieth century.[4] This company was once a regional distributor who bought several other regional distributors to create a national presence. Meanwhile, McKesson purchased General Medical in the 1990s, when General Medical was a hospital and nursing home distributor. In 2006, McKesson sold the Med/Surg distribution operation to Owens & Minor. This sale left McKesson without an acute care distribution presence so that they could focus on their pharmacy and nursing home distribution business segments. Owens & Minor continues to be a strong national player in the acute care distribution market.

## *Medline Industries*

Medline increased its size—a small amount—through the acquisition of product manufacturers, such as Carrington Laboratories and Maxxim. Other than through the purchase of Sun Health, the primary growth and national expansion for Medline has been through the growth of product sales and prime vendor agreements, because Medline is primarily a manufacturer who distributes. Medline, with 2006 sales of roughly $2.5 billion, is the largest privately held manufacturer and distributor of health care products in the United States. Medline has a unique business model that competitors have been trying to emulate for years.

Medline began in its modern form in 1966 by manufacturing several "cost-saving" items to hospitals and nursing homes that were similar to the name-brand products. The manufacturing and contract-manufacturing capabilities of Medline increased through the years, and when the business passed $1 billion in annual sales, it became obvious that total-package distribution was inevitable. Through many unique corporate, cost-savings agreements, Medline's distribution of all medical supplies increased dramatically. Today, in a typical distribution agreement, Medline may be selling 50–70 percent non-Medline products, but in many cases the majority of the products may be a Medline-branded product and the rest is competitive product.

## *Nursing Home and Home Care Distribution*

Long-term care distribution is a different animal than the acute care side, but only because of the difference in the customer. The nursing homes still have to buy products, and use them for the patients, just like the hospitals. The nursing homes still have purchasing people, but their role has much less impact. Traditionally, the size of the orders are smaller, the freight charge per order is higher (due to diapers and enteral feeding product, which weighs more), and the total dollars per order is less. If you walk into an 800-bed nursing home, you will be amazed at the difference in the purchasing function compared to an 800-bed, or even a 300-bed, hospital. A hospital will have multiple purchasing people, while the nursing home will stack this job on the back of one person more often than not. Never the less, the long-term care distribution market is still more consolidated than in years past.

The nursing home market used to have several regional competitors and still has more regional players than the acute care market, but this ever-expanding market is still dominated by the national boys at this writing. The primary reason we find more regional distributors to be common in the nursing homes, I believe, is due to an absence of Group Purchasing Organizations (GPOs) in the nursing home market. The GPOs that you will find in the nursing home market are not as tightly contract-compliance driven, and offer less in the way of savings compared to their acute care counterparts. Furthermore, with the consolidation of some of the "big boys" many owner-operators of nursing homes feel neglected. The small, regional distributors do a great job at making those "mom and pop" homes feel important. Consolidation was inevitable, and will continue well into the future, with more of the "big boys" gobbling up the small operations. The following explains how some of the original consolidation in the nursing home market came to be.

## *McKesson*

Randolph Medical was a regional nursing home sales company based in Chicago. During the early 1990s, General Medical bought Randolph to gain a higher presence in the nursing home market. At the same time, they also purchased Titus and Harris Medical. McKesson then bought General Medical a couple of years later, followed by McKesson's purchase of RedLine, who, at the time, was the largest nursing home distributor in the United States. This consolidation made McKesson a major, major player in the long-term care arena. To date, McKesson has continued to concentrate on its core pharmacy distribution business, even as the nursing home sales expanded. The 2006 sale of the McKesson hospital division to Owens & Minor effectively removed McKesson from the Med/Surg distribution business, but they still push for a larger presence in the nursing home and home care market. At this writing, sales in the long-term care division of McKesson were approximately $650 million annually.[5]

## *Gulf South (What About Those Southern Boys?)*

Gulf South originally saw the light of day as a regional distributor in the Gulf Coast region (hence the name Gulf South). Gulf South purchased some small regional distributors, but their growth has been primarily through sales expansion, helped by the procurement of the entire purchasing contract for Beverly Health Care/Golden Living, the nation's largest nursing home chain. Gulf South then formed a partnership/merger/acquisition (depending on whom you ask) with PSS, who had been primarily a physician sales and service company. As of 2006, Gulf South, with a relatively small outside sales force, lost the Beverly/Golden Living business. This major blow drastically reduced the size of Gulf South's sales in the long-term care market, placing them a distant third in the long-term care distribution scheme nationally and probably behind a few regional players.

## *Medline Industries, Inc.*

As previously mentioned, Medline's growth in the long-term care market has been primarily through adding sales people and expanding sales of Medline product to nursing homes and home care agencies. At the time of this writing, Medline Industries, Inc., is the largest supplier to nursing homes, with sales in excess of $750 million, as well as the nation's largest supplier to the fast-growing home health care market. Medline's growth in the nursing home market has been largely the result of agreements signed with regional and national nursing home chains. In the home care market, the growth is primarily due to the success of their direct patient sales programs through larger home health groups. At the time of this writing, Medline took over the entire distribution agreement for Beverly/Golden Living. Medline is *the* market leader in signing agreements with small, medium, and even large nursing home chains. These larger groups pool their buying power and negotiate better contract pricing for themselves. A company like Medline can be more aggressive, because a larger percentage of what a nursing home chain will buy will be direct from the manufacturer. The company makes money, the nursing homes save money, and the patients benefit by a more consistent approach to care through the value-added services provided by the company.

## *Why Is Distribution Important to Me?*

If you are not working for a distributor, it is vitally important that you know who your customer uses for their prime vendor and their secondary vendors. Additionally, it helps to be fully aware of which distributors stock your products. If the hospital or nursing home uses Distributor A, and that distributor does not carry your product, then you may be facing a larger uphill battle than you know. Conversely, if Distributor A carries your product and has it in stock, some of the work has already been completed for you. Even though the primary purpose of a distributor is to move boxes from point A to point B, their strong relationships in a particular facility can have an effect on the outcome of your sales efforts. I encourage you to build relationships with all the distributor reps for your accounts. It can only help you.

Most of the larger distributors, especially in the hospital market, have become resellers of their own private label products. In some cases, the distributor actually makes, or appears to make, the product they are selling under their own name, but most distributors that sell private labels use contract manufacturers who make the product and then put the distributor's name on the outside of the package. The same manufacturer may make products for several different companies. The idea here is that the distributor can usually save the customer money on commodity-type items and increase their own profit margin over the "name brand." If this sounds familiar, this is an example of companies trying to emulate the Medline model.

## *Speaking of Distribution . . . Cost-Plus and the Big Deception*

In the 1980s, someone came up with the cost-plus program of pricing their products from a distributor to a hospital. The idea seemed sound: The distributor would guarantee the hospital that they would move all those boxes from manufacturer to the account at the cost the distributor paid plus a markup, or acquisition fee. In the old days, a 9 percent distribution fee was pretty good. In today's market, hospitals that operate on a cost-plus distribution agreement sometimes have as low as a 3.25 percent cost-plus agreement. This means that if the box costs $100, the distributor will mark that box up 3.5 percent to get it to the hospital. Sounds good, right? Here's the deal, and I did not figure this out until I had been at this for ten years: I always knew when a cost-plus distribution deal had been signed during my days working for a manufacturer because the distributor would call me and want a price increase on ten or twenty items the hospital was buying. I thought this was strange at first, but then I realized that 5 percent of $100 was less than 5 percent of $120. See what I mean? This is just an editorial, but with cost-plus pricing there is no real incentive for a distributor to save the hospital money on the products, unless the hospital has some other kind of cost-saving or cost-containment contract in place, which often happens. With cost-plus the hospital thinks they are winning, because they have a "COST PLUS 4 percent"

HAVE I GOT A DEAL FOR YOU!

agreement. Yep—look how cool I am—I negotiated this great deal for my hospital. In actuality, a cost-plus program will rarely save a hospital money by itself, unless they were being royally taken advantage of by their previous distributor, which is entirely possible.

This is somewhat off the beaten path but worth mentioning in order to give you a frame of reference for the impact, or lack of impact, felt by cost-plus pricing: In this day of modern reimbursement for hospitals, hospitals are known to operate on profit margins of 1–3 percent. Most people would agree that a slim margin like this does not leave much room for error. Any kind of major build-out or new construction can throw the operating margin into the negative. For a hospital to realize the same $500,000 cost savings on a distribution fee decrease, they would have to increase revenues by $25 million. Wow! Increasing revenues by $25 million is a lot harder than finding $500,000 in savings. The problem, however, is finding a *true* $500,000 in savings. If a typical 300-bed hospital is spending $30 million each year and $25 million is through their current distributor at a 4 percent cost-plus pricing structure, the distribution fee is costing them $1 million annually. If they switch their distributor to a company offering 3.5 percent, they are only saving $150,000, which is not a big savings toward the bottom line. If the 3.5 percent distributor gets a 1 percent price increase from all their manufacturers, they have already increased prices by $250,000, negating the $150,000 savings. The only way to counter this result is to pick a distributor who offers a guaranteed savings program through product and system conversions.

# Chapter Seven

## *Group Purchasing Organizations*

Group Purchasing Organizations, also known as GPOs, have only been around for about twenty years. Back in the early 1980s, a group of purchasing people got together in an airport (as the story goes) and decided that their facilities combined could have more purchasing power if they pooled their collective buying clout under one umbrella. This first GPO was called VHA. Believe it or not, VHA once stood for Voluntary Hospital Association. I say believe it or not because today's VHA facilities, who are part of the larger umbrella known as Novation, are fairly firm about staying with their contracted vendors, and they would not like to hear the "V" referred to as voluntary. Anyway, GPOs were created so that individual hospitals could have more buying and bargaining power in the market.

Think of it this way, let's assume I am a 200-bed hospital, and I will be purchasing $25,000 annually of a particular type of product. Let us also assume that there are ten different manufacturers of products like this in the marketplace, and they have the market split ten different ways. If my 200-bed facility agrees to limit the product choice to only two vendors along with fifty other 200-bed facilities, my hospital now has more power to negotiate prices. From the vendor's standpoint, they know that they will now get a MUCH larger share of that market if they only have to compete against one company rather than nine. As a vendor/manufacturer, if I know that I am going to get close to 50 percent of this group's purchases, I can afford to lower my price in anticipation of much higher guaranteed volumes. If I am the 200-bed hospital, I used to pay $100 per box for that item for which I was spending $25,000 annually, but now since the hospital is part of a much larger group, the price may drop to $85 per box. Power in numbers!

From that day twenty years ago, the number of GPOs has grown larger and larger, until a couple of years ago, when we started to see consolidation of groups. From two or three, the market grew to in access of thirty different GPOs, but now the list is much smaller. GPOs can be divided into three or four basic types: investor-owned (also known as proprietary), alliances, voluntary, and religious-based groups. The religious-based groups usually belong to one of the previous three designations, but some are big enough to start to do their own contracting.

Investor-owned groups may be public or private, but they offer the greatest degree of contract compliance. In addition, their business focus is a little bit different. Investor-owned groups tend to care about things like profit and return on investment, and they generally care more about saving money rather than raking in administration fees.

The alliances are made up of hospital systems as well as individual hospitals, but they usually share some sort of common interest or concern. Many of the members of these alliances saw the benefit of becoming part of a larger group in order to maximize their buying power. Members of the alliances will tend to focus more on contract compliance than a voluntary group, and some are nearly 100 percent compliant.

Voluntary groups are just like the moniker implies: Voluntary! Sometimes they comply, and sometimes they don't, depending on what they want to do at that point in time. The hospitals that belong to a voluntary group will often times belong to more than one group at the same time. Mostly smaller groups and hospitals tend to make up the bulk of the voluntary members, and the only two that really have any impact in our current market are Amerinet and MedAssets.

The following is a list of the major hospital-based GPOs with a brief description and history:

## *Novation*

Novation, which is classified as an alliance, was established in 1998 by combining the members of VHA, University Hospital Consortium (UHC), and Health Professions Partnership Initiative (HPPI) (now called Provista), and now is the contract services organization for 44 percent of the nation's staffed beds, which represents 49 percent of all hospital admissions, according to Novation.[6] All three of the entity groups now have their contracts negotiated and administered through Novation. Currently, Novation is comprised of roughly 10,000 members in the United States, consisting of 200 "shareholders." Novation facilities are fairly compliant to their contract portfolio and contracted vendors. With Novation's buying power and experience, they have been successful at creating unique contracting programs that can help facilities save even more money for maintaining contract compliance across a range of product categories. Novation and Premier are the two largest GPOs in the United States, and together these two perform the contracting function for half the hospitals in America.

## *Premier*

Premier, the other big alliance GPO, was one of the earliest GPOs and currently has over 200 members, or shareholders. This purchasing organization represents over 1,500 hospitals and over 42,000 other health care facilities. Premier was started in Chicago by a few members, including the Jewish Hospital Consortium, who banded together to make a larger group of hospitals that would become shareholders of Premier. Not all Premier members are shareholders, but if they are a shareholder, they seem to be more compliant, but Premier is still what I would call a highly contract-compliant group.

## *Health Trust Purchasing Group (HPG)*

Ughh! This one is a little complicated. Several years ago, Columbia–Hospital Corporation of America (HCA) owned a bunch of hospitals in various parts of the country. As a group, HCA had great buying power, and therefore, some great contracts with vendors. HCA would be considered investor-owned or proprietary. Since HCA owned all their facilities, they could guarantee compliance levels of nearly 100 percent. Someone at HCA had the good idea of forming a new group to market their contracts to other hospitals, so they created the HPG entity and sold the contracting services to other hospitals and groups. Some large hospital groups now covered under the HPG umbrella include: Triad, Life Point, Quorum, HMA, and of course HCA. Overall, HPG facilities are very compliant, but the non-HCA hospitals are not as compliant as the HCA-owned facilities. For the HCA-owned facilities, HCA even has their own warehouses that do the shipping to the facilities, which helps even more to ensure compliance. The newest large member to join HPG is a group called Consorta.

## *Consorta*

Consorta, up until mid-2007, acted as its own contracting organization made up of various groups of hospitals. The largest member of Consorta was a religious-based group called Ascension. Ascension happens to be a group of Catholic-owned facilities. Ascension switched to Broadlane in 2007, which was the beginning of the end for Consorta. With the loss of so many hospitals, Consorta decided to join HPG, so the remaining Consorta facilities started to buy off the HPG contracts, and Consorta is basically no more.

## *Broadlane*

Broadlane's story is similar to HCA/HPG. Tenet owns a large chunk of hospitals primarily in the Southeast and the West, and they had good compliance and great buying power, much like HCA. Two gentlemen who worked for Premier approached Tenet about creating a new GPO to sell the Tenet contracts. So Tenet created a purchasing/contracting group called Broadlane in order to sell their contracts to non-Tenet facilities. This GPO has grown fast, and now Tenet makes up only about 20 percent of the shareholders of Broadlane. Broadlane/Tenet includes groups like Continuum, Kindred, Universal, and the Kaiser system. Since Ascension left Consorta and decided to join Broadlane, the size of Broadlane increased by 20 percent. This makes the buying power of Broadlane now even stronger.

## *HealthSouth*

You may have heard about this one in the news, or at least you might have heard about the troubles their CEO has been through in the last couple of years. HealthSouth owns all their facilities, which are largely dominated by surgery centers, so they would be considered in the category of investor-owned or proprietary. In addition to the surgery centers, HealthSouth has a number of rehab centers they own, which are like high-level care nursing homes. Since HealthSouth owns all their facilities, their contract compliance is very high.

## *Amerinet*

Amerinet, which falls into the voluntary category of GPOs, is what some people like to call a *secondary* contracting organization. Many hospitals that buy off the Amerinet agreement have another GPO they use for their primary contracting entity, as I mentioned earlier. However, some of the hospitals that buy off the Amerinet agreement are in fact using Amerinet as their primary GPO, like Inter-mountain Health Care (IHC), Amerinet compliance can be quite spotty, and most facilities will use the contract if it helps them, and not use the contract if they do not need it.

## *MedAssets*

MedAssets is a unique GPO. MedAssets sells their contracting services to facilities, and the facilities that use them are compliant. MedAssets' contracting is what separates them from other GPOs. MedAssets will approach a hospital and guarantee cost savings without changing vendors many times, and they market themselves as more of a technology company. If the hospital signs on with MedAssets, MedAssets will then negotiate with the hospital's current vendors to get a better price based on higher levels of compliance. MedAssets tells potential members that they focus less on national contracting and focus more on customized contracting for the individual hospital or Integrated Delivery Network (IDN). Even though hospitals with a MedAssets contract are relatively compliant, MedAssets often offers little in the way of contract commitment to the supplier. MedAssets may be the fastest growing of the GPOs over the last couple of years.

## *Religious-Based Organizations*

I mentioned Ascension earlier as a religious-based group, and this one really stands out because of its size. Ascension is the largest religious-based organization, and they currently have negotiated over 140 of their own contracts utilized in 75 facilities. Ascension's 2006 operating revenues were $11.4 billion,[7] which is why you can see what effect their decision to leave had on Consorta. Other religious-based groups include Catholic Health Initiatives (CHI), Sisters of Mercy, Catholic Health West, and Sisters of St. Joseph of Carondelet, which is now part of Ascension. These religious-based groups trend toward higher levels of contract compliance, but their focus is more on non-business issues. As the reader can imagine, religious-based organizations are not as concerned about profit as other groups, but they do want to try to maintain a healthy bottom line so that they can continue to provide care to the indigent population. A religious-based hospital will rarely, if ever, turn away a patient for lack of money or insurance.

*****

At this point the reader may be asking, why is a knowledge of GPOs important, and what does it mean to me as I sell my products or services? You probably figured out that you need to know two things: To which GPO does this hospital belong, and are my products on contract with that particular group? You absolutely need to know these two things. Early in my career, the GPO status was not that important. In today's market, however, almost all hospitals are compliant and reliant on their group purchasing contracts. In the majority of hospitals, Materials does not even want reps showing products that are not "on contract," which is a change from years past. Does this mean that hospitals do not buy off contract? Of course not. If my company has a unique, revolutionary, proprietary product with limited or no competition, getting the product onto a group contract does nothing for the company. Why would the company want to lower their price to increase compliance when they are the only game in town? In a situation like this, most companies will charge whatever they can get for as long as they can get it. Once competitors enter the marketplace, their perspective will change, and then maybe they will need to establish a contract with the groups.

I teach my reps to always know the GPO status and information and always carry a copy of the contract pricing with them on every sales call. Believe me, the question about contracts and group pricing will come up, and if you have the contract with you, you will be better prepared to answer those questions and look more professional. I cannot stress enough the importance of knowing not only which GPO the hospital uses but also whether that GPO is committed or non-committed. Some hospitals, like HCA and Tenet owned, are going to do whatever their corporate decision makers want them to do with regard to contracts and contract compliance, for the most part. If a hospital belongs to Broadlane but is not Tenet, they may decide to do something entirely different than their Broadlane brethren. If a particular hospital is part of a system with shareholder status of that group, chances are that they will be more contract compliant than a non-shareholder facility.

I cannot stress enough the importance of understanding and knowing the national account affiliation of your facility. Congress and the Justice Department called Novation and Premier onto the carpet in 2004–05 because of some of the feedback they had from small manufacturers who claimed the hospitals were being deprived of life-saving technologies, because of the failure of those small medical device manufacturers to be on contract for various GPOs. Despite this fact, the GPOs seem to have more power than ever on the decision-making status of the hospitals. If you are on contract, there is great opportunity there. If you are not, maybe you can get in under the 10 or 20 percent area of non-compliant vendors. One other thing to consider is the power of large hospital groups in one geographic area, usually referred to as an IDN. When a group of hospitals in one city has a large presence in that market, such as BJC in St. Louis, which owns over fifteen hospitals, the IDN can have some buying power all by itself. As a matter of fact, many IDNs are starting to rely more on their own contracting people to negotiate deals directly with the manufacturers and suppliers.

Most contracts require a compliance rate of 80 or 90 percent to get the rebates from manufacturers. Oh, did I fail to mention that part? These GPOs usually get fees for contract compliance and purchases from the manufacturers, and the hospitals usually get rebate fees for contract compliance. That is all part of the savings. If you work for a company who has a contract with, say, Premier, and if that facility buys $100,000 of that contracted item, then Premier may get up to 3 percent of the sales total. This is how Premier and other GPOs keep the doors open. If the reader works for a company who is not on contract, you still have that 10 to 20 percent area where you can get the facility to buy your products. Here is the other key: Most hospitals will sign some sort of letter of compliance (LOC). On that LOC, they will list their expected annual spending in that category. For instance, if the hospital buys $500,000 worth of sterile procedure trays in a year, they will indicate that in their LOC at $400,000 expected spend ($500,000 × 80% = $400,000). Here is a news flash: The hospitals almost always under-estimate their spending, because they want to make sure they get *their* rebates. Oh yeah, I forgot to mention that too. The hospital will get a portion of that 3 percent. So, if the total rebate is $3,000, the hospital might get $1,000–$2,000 if they comply with the contract. This may not seem like much, but if you multiply this by thirty different contracts and an annual contracted spend of $15 million, the rebates add up.

Where is the opportunity you may ask? If the hospital says they spend $500,000 annually on a particular product, they may actually spend $650,000. If they have an LOC for $400,000 ($500,000 × 80% compliance), and they really spend $650,000, then they could realistically buy $250,000 worth of your products and still be compliant on their contract. Bam!!! Look again. The level of spend is listed as $500,000. Eighty percent of that is $400,000. They actually spend $650,000 on that category. Six hundred and fifty thousand dollars minus $400,000 leaves a non-contract expenditure of $250,000. Wooh-hooh! Rack it up baby! I can sell $250,000 worth of my product to this hospital and NOT interfere with their contracted rebates and status. I don't know about you, but that sounds like opportunity to me.

Okay, but maybe you, the reader, are thinking that your new job does not take you into hospitals, it only takes you into nursing homes and/or home health care agencies. You may be thinking, are there GPOs for nursing homes and home care? Yes, but the compliance and impact is far, far less. Remember Novation? Novation has a sub-group under their contract called Provista (formerly HPPI). Provista has quite a few nursing homes and home care. If that particular facility believes in utilizing their national agreements, this may provide some kind of an advantage for you if you are on that agreement. To be honest, the nursing homes and home care agencies seem to do what they want with regards to which products and services they utilize. What is far more important is the membership status of that nursing home or home care agency. Huh? Yes. Many nursing homes and home care agencies are part of a larger group that not only tells them what to buy but also from whom to buy it.

If a nursing home or home care agency is part of a national group like Gentive/Healthfield, Beverly/Golden Living, Sava, SunBridge, Interim, etc., the corporate ownership office may have a VERY large part in determining which products and suppliers that facility uses. In addition, the national group may use a central purchasing provider, like DSSI or Serus. These purchasing providers act as a clearinghouse for all the purchasing functions for that group. Why? Because these entities can provide accurate tracking data on the group and facility's spend on categories and products. What does that mean to you? Your products will need to be in that system. Remember, however, if you are selling to nursing homes or home care agencies, national agreements are, by and large, much less important than at the hospital level in most cases.

Does all that seem complicated? It is not. Your company will give you clear information related to the status of your products and your company related to national account, or GPO, status. Keep in mind, however, if you are unlucky enough to be with a small company with no GPO agreements, or if you happen to be with a big company that has not been able to sign any agreements for the products you will be selling, then you still have that area of 10 to 20 percent plus, if your product is that much better than the competition. Or if the clinician wants it that much more, they can usually find a way to get it in, because ultimately the health care facility wants to provide quality patient care.

This page shows a good breakdown of the groups, number of members, and their approximate 2006 purchasing volume.*

United States Major Group Purchasing Organizations[8]

| **Group** | **Number of Members** | **Annual Purchasing Volume** |
|---|---|---|
| Novation | 1,370 | $17.0 Billion |
| Premier | 1,800 | $14.0 Billion |
| MedAssets | 1,255 | $10.0 Billion |
| HPG (Consorta incl) | 994 | $ 7.0 Billion |
| Broadlane | 525 | $ 6.5 Billion |
| Amerinet | 1,530 | $ 4.7 Billion |

* Figures based on HHM research data

# Chapter Eight

## *Sales Basics You Need to Know*

If your only exposure to selling was the guy who sold your family their first Electrolux vacuum cleaner, or the guy at the local new or used car lot, you have the wrong idea of sales, and especially of professional selling. Nobody—well maybe I should say almost nobody—sells that way to the health care buyer. Unfortunately, I still hear stories from customers about so-and-so who "seems like a used care salesman." You know who you are!!!

For the most part, things are not like that anymore, at least for successful sales people. Today, we are referred to as "sales professionals," or "sales specialists." The health care market has no room for anything but sales professionals. We are selling products that are used on patients, which can literally have an effect on whether the patient lives or dies. The last thing a clinician needs when buying today's modern products and services is a guy named Slick who is trying to cram something down the buyer's throat. This is just not acceptable.

## Old-Time Selling

Some fairly famous authors, lecturers, and managers have been quoted as saying, "Nothing happens until somebody sells something." When I took my first marketing class in college, the class was taught by a gentleman who actually used to call on my father when he was the director of Materials Management. This professor was excellent, and I learned a great deal from him, but when he asked the question about this old sales adage, he said that sales cannot happen until marketing has studied the market. Thus began the first of many disagreements on this subject between teacher and pupil.

I realize I am digressing just a tad, but there will be a point to it, so don't go away. Some truth exists to what my professor was saying, but my argument was this: Someone had to sell somebody at the company that somehow it was worth studying *that* market to see if we should SELL SOMETHING! As you can plainly see, I was right, and he was wrong. Regardless, the "nothing happens until someone sells something" could be considered a philosophy of "old-time selling," even though it still applies today. If you had any doubt, I can tell you that I heard noted sales trainer Bryan Tracy state, "In order for anything to get started, someone has to sell something." I also heard Steve Forbes, of *Forbes Magazine,* say the same thing, believe it or not.

In the old, old days, sales people were described with adjectives like slick, smooth, could charm the skin off a snake, can talk anyone into anything, carpetbagger, etc. Old-time selling was synonymous with scamming or talking people into things they did not need, like swampland in Florida. In the days of westward expansion, sales people would come to town selling everything from snake oil to advanced gizmos for building log cabins. A salesperson would come to town with his wares, place himself in the center of town, and after rallying a crowd together, start pitching his product to the masses. This person had one shot to sell his product, and one shot only. He had to be slick, convincing, and, most importantly, believable. Then the automobile was invented, and car salesmen took it to a whole new level.

This was long before the Food and Drug Administration, as you can imagine. As a matter of fact, this is probably why we have the FDA. Today's market is much different whether a sales rep is selling copiers or medical supplies. However, there are some basics that one has to master in order to be trusted and successful. The bottom line, however, is that old-time selling is out the window, in general, and especially in health care.

## *Sales Basics for Any Market*

If I had a nickel for every time a friend or acquaintance said to me, "Oh, you should meet my third cousin Billy Bob Joe. He would be great in sales. He can talk your ear off," then I would have a lot of nickels. (I bet you thought I was going to say, I would be a rich man? Let's face it; people can't really get rich on nickels. This is the twenty-first century.) Even though Billy Bob Joe can "talk your ear off," that does not mean that he will be successful in sales. As a matter of fact, he will probably stink in sales, and even if he found a sales job, he would not find many customers wanting him to come back for a second visit. Despite what you may have heard—and this is aimed at the clinical, or "non-sellers," reading this book—sales is all about listening, not talking. But when you do talk, you better say the right things.

In my Preface, I talked about attending a sales training program sponsored by the wonderful folks at General Electric. At the time, this sales training program was called the "Results Oriented Approach" to selling, or ROA. I was twenty-three or twenty-four when I attended that program, and the basics I learned there are still with me today. The GE training and the training by many other sales professionals center on the concept that the more you know about your customer, their business, their products, likes and dislikes, along with what they truly want, the better you will be able to help them. This approach is sometimes referred to as "needs oriented," or "problem-solving." The sales rep is trying to solve problems by addressing the needs of the customers with the rep's products or services. I am not going to go into any tremendous detail on sale techniques, because there are a multitude of fabulous books on the subject, and because I want you to understand how to sell to the health care customer.

After having said that, I know that some sales basics are important to master in order to be successful in health care sales. The following is a breakdown of some of the important basic things you need to know and do:

## #1 (And Most Important) Asking Questions!!!!!!!

Why is asking questions important? I thought I really needed to know how to sell my PRODUCT? Well, how are you supposed to know what your customer wants and needs if you don't ask the questions? Do not assume things with any of your customers. Even in your excitement, you need to go back to the basics of asking questions. Quick little story here: Early in my sales career, I started selling an amazing product that was revolutionary for the wound care market. Everywhere I took this product it made a huge difference in the way wounds healed. I cannot begin to tell you the number of patients with serious wound issues that experienced life-changing results with these products. Anyway, I was so excited to share this product with my customers, I would come into a one-on-one meeting, and after establishing a rapport I would launch into my well-polished sales presentation with all my technical information and just wait for the explosion of excitement from my customers.

I was calling on one particular ET nurse, who at the time was my primary call point because the ETs were the nurses in the hospital that treated the wounds. Refer back to my chapter on politics for a refresher on the ET nurse. Anyway, I was at a hospital in St. Louis, and I sat down with one of the aforementioned ET nurses and launched into my exciting sales presentation about how these new products would revolutionize the way she treated wounds. When I was finished, I asked the nurse what she thought about the products, and she said, and I quote (because it still stands out in my mind of what a major mistake I made): "It sounds great, but I don't do wound care, I only see the Ostomy patients." You've gotta be kidding me!! I had just wasted the last twenty minutes talking, talking, and talking without asking one single question. We have all heard of the old 80/20 rule, which can mean many things, but in a sales call, the 80/20 rule refers to the amount of time you should spend listening compared to the amount of time talking. I had it reversed, but I never did that again . . . ever. Had I first asked the question, "What kind of wounds do you get involved in treating?" I could have saved us both a lot of time. Keep in mind, in a one-hour sales call, following the 80/20 rule still gives you twelve minutes to talk, which is a lot. Even if I am working a trade show with very little time to talk to each customer, I still start by asking them questions about what they use and why.

You may have a new product that you want to show the world. You may have just completed your product training on this product, and you are convinced this new product will save the world, and as a rep, you want to shout all the product attributes from the rooftops. Well, "Lighten up, Francis." (A little quote from the late Warren Oats in the movie *Stripes*.) Slow down; easy does it. Does your customer see patients who have the conditions necessary to use this product? Is that customer who is wearing the dollar signs in front of you the key decision maker? Is this customer using a similar product now? Who else in the facility uses this type of product or sees the type of condition for which this product is indicated? What types of things are important to the clinician when selecting a product to use on patients like this? What is the process for trialing or buying this product? If the customer is using something similar, what do they like about it? You must ask questions! Sales professionals in any business are taught to try to make their questions open ended, but sometimes you have to ask closed-end questions. Closed-end questions usually come at the end of a series of questions. Closed-end questions are fine, and sometimes those are the questions that lock the customer into a certain commitment.

If you know anything about psychology, you know that the person who is in control is not the person talking but the person who is asking the questions. If you are good in sales, you should be able to ask questions that lead your customer right where you want them. Even though your customer is doing most of the talking, you are passively in complete control of the sales call as long as you are asking the questions. Think about it: No matter what the customer says to me, I will take that conversation right in the direction I desire by simply asking the next question.

Let's just say that you have a difficult customer who does not really like your company, and he or she did not really like your predecessor. You are presenting your product and asking questions, and the customer says something like, "Your company's products suck! I had a widget from you guys that broke in half." Now, if she is in control of the conversation, you are going to spend the next ten minutes talking about that fifteen-year-old widget. Think about this: What if the first thing out of your mouth is, "Mary, I am truly sorry about that. That should not have happened. Is the quality of the product the most important thing to you when selecting what you are going to use in this facility?" Wham! The rep is BACK in control. Just like that. If Mary tells you that quality is the most important thing, then you will probably ask her something like, "So Mary, if the quality of my product is there, price doesn't really matter, correct?" Even though this last question is closed-ended, it pins her down. As I mentioned in the earlier paragraph, the closed-end question here establishes that Mary will put quality above price, so if my product is of the highest quality and meets her needs and fulfills her wishes, she better not come back and say it is too expensive, because I locked her in on the quality issue. Anyway, always ask questions.

## #2 *Know Your Customer*

This is an area where so many reps fall short. Knowing your customer starts with simple things, like you will not be calling on an urologist for spinal column stabilizers, just like you will not sell scales to the Radiology Department. Make sure that you are at least in the ballpark of products before you even try to get an appointment with your customer.

Much of the knowledge a medical sales rep will

gain about the customer will come in the one-on-one sales call through the proper use of asking questions, but many other things can be learned before the new rep even enters the customer's office. Let's just say you have an appointment with the DON at a large nursing home in your area. You sell seismic widget adapters, and you know that every nursing home should use them on all their patients if they want to deliver quality care. If this were my appointment, sometime before that day I would have made some phone calls to find out which distributor the home uses, and I would have talked to the distributor rep for that account. The night before, I would have gone online to check the number of beds in the home, which tells me if they are large or small. I would also have gone to the Medicare.gov website and checked the status of the home to see if they had received citations that my adapter may be useful in helping to curb. In addition, when I get to the home, I am going to ask the receptionist if the DON has been really busy and if she is in a good mood. Being really nice to the people that answer the phone and work the reception desk will pay large dividends. I think you get the idea.

## #3 Know Your Competition

Is there any other company out there that makes a seismic widget adapter? Hopefully, your company covered competition when they trained you on your new product, but even if they did, the learning should not stop. Visit your competitor's website. Call their customer service line. How were you treated? Request some literature be sent to you. Read the info. Does it make sense? Do you see any opportunities for your product?

Get some samples of your competitor's product. Play with the samples. Use them as they are intended to be used. In your opinion, how does the product stack up? Ask your co-workers how they think the product stacks up to yours, and ask the "experts" at your own home office. If you find a customer who uses your competitor's product, ask them questions about that product, such as likes and dislikes, price, quality, etc. Your customers will usually be very honest about their likes and dislikes of competitive products, so much so that they may help you sell your product to themselves.

## #4 Know Your Product

This should be a given, but there are many degrees of "knowing your product." Some reps go to the sales training class for the new product, and they listen intently, and they read over the brochure, and now they are ready to go sell it, right? Wrong! Read the literature again and again. Read ALL the clinical information associated with your product. Play with your product. Use it. Try to take it apart. You need to know the product like the back of your hand, because the product is an extension of you. People sometimes make fun of me because I really use all our products and I digest all the clinical information, but I can say with confidence that I am almost never left with a dumb look on my face during a sales call or a "I will get back to you on that" reaction when I am asked something specific about my product or the competitor. I cannot tell you how many times a question has come up from a customer that related to clinical information I managed to absorb and recall because I took the time to study and learn. Not only does it make me look like I know my product, but it reflects well on how my company trained me. When I know my stuff when it comes to my product, and I am confident with it in front of my customer, and I can talk about it compared to what they currently use, then I automatically bring more value to the relationship. In addition, having this knowledge, whether I use it or not, gives me the confidence to make a better presentation.

## #5 Sell Yourself, Sell Your Company, Sell Your Product

Regardless of what product or service you are selling in the medical arena, you need to always consider every customer as a potential sale of not just this product but many products to come. Today you may be selling blood pressure cuffs, but tomorrow you could be selling sterile procedure trays. Think of your customer as the entire facility, not nurse Mary at Deaconess Hospital, but Deaconess Hospital in its entirety. We have all heard the sales adage, "Sell yourself, sell your company, and sell your product." Notice that the product that you are so excited about selling is last. If you sell yourself by coming across as a good listener, addressing the customer's needs, making the customer feel good about your company, and knowing your product, then you will be able to see that customer at some time in the future about some other product. It is true that people still like to buy from people *they like*.

None of us can say enough about the importance of building relationships with your customers and potential customers. The more people who know you, like you, and respect you, the larger your potential market becomes. Spend time with your customers. Bringing in lunch for an "in-service" does very little to build relationships, which is why I have such a problem with the pharmaceutical reps. A big part of their job is bringing lunch in for the entire staff to spend two minutes with the doctor. Try to get them out of the facility to lunch, dinner, or breakfast. If your company does not allow entertaining (which will probably start to apply to most of us), no one can keep you from going to your customer's son's little league game or their daughter's dance recital. Warning! You have to genuinely like people and develop a genuine interest in your customer and your customer's interests. If this is not something you can do, you may be in the wrong business. From my experience, I find that the people who come from a clinical background and move into medical sales have the most trouble with this concept. If you think about it, this makes no sense what so ever.

Relationships often develop over common bonds or interests. If you come from a clinical background, you stand a better chance of having something in common with your customer, who probably *is* a clinician. Having common bonds with your customer can only help you. Use what you have! Play up the potential clinical bond. Some of my reps who can sell, who are also nurses, can tell anecdotal stories from the perspective of a nurse, and this has more credibility because they are calling on mostly nurses. If you are calling on a doctor and you do not know the interests of that physician, ask someone in his office. Does he play golf? Does he have kids? Does he like fishing? Or simply ask the receptionist what the doctor likes or does in his spare time. Believe me, she will know. The nicer you are to her, the better chance you have of finding that important, but sometimes elusive, information you desire.

# Chapter Nine

## *Technical Selling*

Some products are more complicated than others. Some markets are more detailed and complex than others. If you find yourself selling products that require some technical expertise, you better do all you can do to develop *that* technical expertise. Your company should give you the tools necessary to gather the knowledge you need for your product, and they will hopefully give you information about the market and your competition, but as I stated in the previous chapter, you still need to learn all you can in order to be truly outstanding and an asset to your customer.

The problem with "technical selling" is the degrees of knowledge your customer possesses or does not possess. If you can go into every sales call with the attitude that your customer is more knowledgeable about the application for your product than you are, you will start out on the right foot. The best way to sell a technical product is to be as un-technical as possible. I know that sounds strange, but try to keep it simple. Even the most complex products and markets can be explained in a simple fashion if you take the time to find that simplicity. If your wife or husband cannot understand what you are telling them about your product, you may be too technical. I know that sounds crazy, but it is true. Yes, your husband or wife may not understand the market and the application, but they should be able to understand the point you are trying to make. When you are talking to your actual customer, you can always get more technical by being prepared to answer questions that will most certainly be asked. Keep in mind, you *should* know your product better than your potential customer, and if you do not, you need to go back to the last paragraph. If you have that confidence, you can deal with most any customer.

Other than the development tools mentioned above, there is a great way to improve your technical knowledge and your technical selling skills that will not cost you a dime. For the most part, nurses and clinicians love to teach. This is the one overwhelming commonality I have found in medical sales. With this in mind, you can use your customer, who is your potential buyer, to make you better at doing your job. How, you may ask? The answer is through preceptorships and work contacts. Since most of our customers truly enjoy educating, let them educate YOU! Ask them if you can spend a half-day with them in the clinic, OR, nursing home, or whatever. Tell them that you will bring in lunch if you can just spend some time learning more about what they do. This really works best when you are new to the job. By the way, being new commonly means being at the job less than eighteen months. Some people get away with this "new" business for two years, and other can only pull it off for a year. Either way, use what you have!

Some of my most important and impactful moments in my sales career have been during times that I have been learning from my customer or potential customer.

When I started in wound care sales, I knew very little, and I had a seven-page guide to teach me all I ever wanted to know about wound and skin care. My mother had some training in this area, and she was a great help, but I still needed more. My mom hooked me up with some ET nurses in our area that took me under their wing and taught me the ins and outs of wound and skin care—at least what was important in 1986. I will forever be grateful to Jan Jester, RN, ET for all she taught me.

By the early 1990s, I was pretty much an expert in my area of sales, or so people thought. I met a doctor, nurse practitioner, and physician's assistant at a teaching hospital in my hometown, and my world was really about to expand. For some reason, these guys let me come into their clinics every other week to observe what they were doing with their patients and some of our products. This was an incredibly valuable learning experience, and it did not cost me a penny. I would show up early in the morning. I would help them do whatever they needed to get the day started, whether it was emptying trash cans or grabbing paperwork, and patients would start rolling in around 8:00 a.m. I stayed each day, and I was there as long as I could or as long as they needed help, which was usually until one or two in the afternoon. I cannot begin to tell you what kind of unique stuff I saw there. You cannot buy this kind of experience, and you can find it in your area too. Talk with your customers/clinicians, and let them know that you are eager to learn. Believe me, most of them will be anxious to teach you.

Thank you again to Dr. H, Bill L., and Owen.

## *Pitfalls of Technical Selling*

*(How to Recognize the "Expert" in the Crowd, and How Not to Perturb Them)*

Some clinical people do not like sales people, no matter how nice and friendly you are and despite how well you know your product, the application, and the market. This type of person, who normally sees themselves as the "expert," can be difficult, but only if you let them. Generally, I believe they are insecure and probably know less than they think, but you cannot have that mindset. First, you need to understand how to recognize them, and recognize them quickly. The good news is that the "expert" will usually let you know right up front that they are the "expert" and you are nothing but a snake oil salesman. How you handle this person may determine your entire future in that particular health care facility, so don't blow it!

There are two basic ways to approach the "expert." The first way is to hit them head-on with all your facts and data to back up what you have to say and to thwart what they are saying about you or your competitor. If you are never coming back to that particular facility, this is the most fun way to go, as long as you can pull it off. By the way, if you truly DO NOT know more than that person, which is usually the case, you better bale right now. The second way is to encourage the "expert" to teach. I have found that the level of information the "expert" thinks they know is directly proportionate to the amount of educating they want to do. LET THEM EDUCATE YOU!!

If you meet someone who has you massively outgunned, go with it. In a non-patronizing way, let them know that you would love to learn more from them and that you would love to see some of the things they see. Oftentimes, this type of person is a pain to start, but very willing to help you—"the idiot"—learn more about what it is you need to know. I pretty much guarantee that if you ingratiate yourself with this person by letting them teach you "all they know," they will become one of your best and most loyal customers. I know this all sounds a bit manipulative and narcissistic, but that is because it is manipulative and narcissistic. What do you want from me? I *am* a sales guy after all.

# Chapter Ten

## *Group Selling to the Committee (Who* Is *Your Buddy?)*

Lets say that you made it by your primary call point, and you somehow made it by the expert, or, God forbid, you have the support of one of the "experts," but your product has to go through the VAT (Value Analysis Team), VAC (Value Analysis Committee), NPC (New Product Committee), PEC (Product Evaluation Committee), or SWAT (Skin and Wound Analysis Team) or ABC, XYZ, as easy as 123, as simple as Doe Rae Me team. Now you may have the opportunity to go before the committee. This is a good thing. If you do not get to speak directly to that particular committee, you have to rely on your customer to present the pros and pros of your product, but I have no problem asking my customer exactly what he or she is going to say to the committee.

Not far removed from the "expert" is the dreaded "committee." Committee-selling is a science in and unto itself, but it can be easy if you have taken the time to make the right friends. I have been involved in selling scenarios where my competitor had previously met, and endeared himself or herself, to every person on that committee. Despite the fact that I had a better product, with better clinical support, and even at a better price, I got my butt kicked. (If I did not mention it before, almost everything I know is learned from doing it wrong the first time but never making the same mistake twice.) Working the committee in a hospital is far more valuable than all the knowledge you have gained about your product, the competition, and the market. If you think I am lying, just try and do it the other way first.

When I train new reps, they always want to ask, "How many samples should I give them, and how long does it take to close the deal?" As far as how long it takes, that really varies, and as far as how many samples you need to give them, that also varies. The thing that is most important for you, the rep, is finding out what the process is in that facility for bringing in new products. Undoubtedly, the process will involve some sort of committee if you are selling to the hospital. Today, every hospital has some kind of committee with an acronym the hospital created as an excuse to get a bunch of clinical people together to argue about what they should or should not bring in to that particular hospital.

The only way to sell to "the committee" is to sell to every individual person on that committee before the committee even meets. Most of your customers will tell you who, or what departments, are on the committee that applies to your product, if you just ask. If you are able to get the names of all the people on the committee, you should be able to contact them. Let this new contact know that Nurse Mary is evaluating the seismic widget and that Mary thought you might be interested in seeing the seismic widget also. The more people you can meet with and schmooze before the committee actually meets, the better your chances of success will be.

Once you are in front of the committee, you need to shine. You and your product are not normally the only things on their agenda, so these people are sometimes together for hours before you get to speak. When you present to the committee, you must be concise and precise! Tell them what you are going to tell them, tell them, tell them what you told them, and then ask for the business or review why they need this product in their facility. Most committees that are made up of nurses are very "touchy-feely," so if you have products that you can actually get in their hands, I would do so, but do not let it detract from your presentation and your message. Something to keep in mind: If you rock during this presentation, you have just hit a home run with a whole bunch of people at one time. Conversely, if you stink, you just lost all those people at one time, so make it good. I have been in medical sales for twenty-two years, and I still practice every presentation I have to do before a group of any size. I will practice by myself, and I will practice in front of my family, and I have even practiced in front of a row of my daughter's stuffed animals, but I still practice, and so should you. We will discuss this more in the chapter covering presentations.

I want to share with you a little bit of insight I have developed over my years of selling to committees. Pay attention to who sits next to whom. The majority of the time the people sitting next to each other are friendly and are on the same wave-length on many issues and levels. Rarely, will you see Mary-Sue sitting next to Norma-Jean, and they hate each other. Don't you remember high school? You sat by your buddy in class if you had the chance. Committees, in my experience, are much the same. You may be asking, okay, how does this help me? Don't be silly! Just as alliances are built on the battlefield, they are built in the health care setting, especially in the hospital. If two committee members are sitting together, it may be likely that they are aligned, and they will like and dislike the same things, people, and companies. Do not minimize this fact.

The question is, how do you take advantage of these alliances? First, if you sell one, you sold them both. If you convince one, you likely convinced their friend as well. I prefer to play to this fact. I try to make eye contact and do much more reading of the people who are aligned but who do not seem to be outwardly in my favor. I monitor my presentation and try to play to those people as much as possible. Also, know that generally one of the aligned people will be the vocal one. You want to address this one, but you want to make a lot of eye contact with his or her friend. For example, let us say that the vocal one is hammering you on a point to which she is misinformed and in the wrong. You should address her point, but frame the last part of your answer with pleading eyes to her buddy. This is a silent way of saying, "Your friend is being a pain in the behind, and she is not correct. Can you help to keep her in check?" I know this may sound strange, but it works.

If you have two people sitting together, and those two people are both on your side, you want to give them some eye and smile time as well, however, you want to share it between them equally, and you cannot afford to give them as much eye and smile time as the people you really need to convince. This is very much like running for political office. A good politician acknowledges and shows appreciation for their supporters, but they want to spend more than half their time convincing the detractors or the people on the fence, that this politician is the right one for them. It just makes sense.

These are some obvious basics, but when selling to a committee, you want to project an air of confidence, and you better know your stuff in order to be confident. If you do not know your stuff well enough to be confident, then do one of three things: 1. Go back to square one and learn, learn, learn—practice, practice, practice. 2. Get someone better than you to do the committee presentation, like your boss. 3. Quit, because you are in the wrong business. I digress. Here are some bullet points to help you:

- Project your voice so that all can hear, but do not shout, and do not stand there like a stiff piece of wood. Move around as much as space will allow, and use your hands to gesture and show impact. You do not have to be the Reverend Jimmy Swaggert, but keep it lively.

- Introduce yourself and your company.

- Tell them what you will be discussing and that they may ask questions.

- Smile as much as comfortably possible without looking like you are the court jester.

- Make eye contact for three to seven seconds on each person and then shift to another person. Try not to go down the row with your eyes. Shift your eye contact around the room, but not too fast.

- Cover your points in a logical order, and even if you are interrupted, go back to where you left off. This is

YOUR presentation—be in control.

• When listening to a question, look directly at that person until you are ready to answer. This shows you are paying attention to them and you care.

• Before you ask for questions, review what you have covered, especially the key points. (I like to do this before questions, because sometimes people stop paying attention when questions are asked, especially if the questions are silly.)

• When you are done, ask for questions. If there were a lot of questions, unless they are clear "buying sign" type questions, you have not done a good job during your presentation of explaining your product and knowing their concerns, but you can still save it.

• At the end, if there were a lot of questions, review the bullet points of your presentation again.

• Ask for the business. Nothing is wrong with letting the members of that committee know that their business is important to you and that they will be happy with your product.

• If I do not clearly know what the next step is, I am not afraid to ask the members of the committee that question. Someone on the committee will usually step up and answer your question.

• Regardless of how they acted, thank them for their time and attention.

When I sell to any committee, as I stated earlier, I always, always practice my presentation the night before, or even a couple of hours before, at least twice. Trust me.

It will make you more confident if you have said these same words in the same way a couple of times before. The only reason not to practice a big presentation is if you do not care about the outcome. Singers, athletes, race car drivers, and actors practice or rehearse, why shouldn't you? It is nothing for a high school basketball player to shoot 100 free throws every day. Can't you spend a couple of hours making your presentation the best it can be? If you do not bring your "A" game to the committee, you will likely fail, and I will personally hunt you down and give you thirty lashes with a wet noodle.

# Chapter Eleven

## *Relationship and Top-Level Selling*

Remember that thing we discussed about people preferring to buy from people they like? Well, it *is* true. Think about it, who is it easier to say no to, one of your good friends or some guy that just knocked on your front door with whom you have never met? The more friends a sales rep has who are customers, the more successful that rep will be selling his or her product to those customers. This fact is without dispute. Building relationships does not happen overnight. I have seen incredibly good reps who were successful without developing relationships, but that is rare. Some of those same reps have had their clocks cleaned by less-formidable opponents with inferior products who took the time to build relationships. In the health care market, unfortunately, this happens more than you might think. I know the reader would like to think that all the doctor's, hospital's, and nurse's decisions were entirely based on the best practice, but it is not always the case.

In the old days, before the government began accusing everyone of fraud and abuse, entertaining was huge in health care. For some companies who have not adopted the PhRMA Corporate Compliance[9] policies, lavish entertaining still is the norm. Some companies have flown doctors and nurses to exotic places to "learn about their product" for a week with their spouse. That is an easy way to build a good relationship, I am sure. There are still other things the rep can do to build relationships, but some of what a medical sales professional does will depend on the policies of the given company and that facility. In addition to companies, many facilities have adopted policies prohibiting any lunches, dinners, gifts, or even a free pen for their employees.

Having said that, many avenues are still wide open to develop relationships with your customer. Many good books are out there that deal specifically with relationship selling, and I would recommend you read one or more of those books to truly learn the ins and outs of selling by building relationships with your customers.

I will give you a few bullet points that seem to work well without spending money. These points, at the end of this section, are assuming that you cannot take your customer out for lunch or dinner or other types of entertaining. If your company and the health care facility does not bar entertaining, by all means, get those folks out of the house, so to speak. When people leave the facility, it takes pressure and distraction away from them, and this gives you an opportunity to learn more about your customer as a person. Remember, people like to buy from people they like, and people LOVE to talk about themselves. Some reps are fun to have out in a group because they have great, funny stories, and there will be a time for that. When you are out with a customer, you want the focus to be on them. Remember the 80/20 rule? You talk 20 percent of the time and your customer talks 80 percent of the time. The same goes for entertaining. Ask your customer questions about themselves, their interests, family, high school, whatever. I am genuinely interested in people's lives, and I love to ask them to take me back to when they were a kid and tell me their story from there. Get them talking! Even people that are not much for conversation love to talk about themselves and their family.

- Do they have kids?
- Do their kids play sports? If so, go watch one of their kid's games. It does not cost you anything, but the goodwill goes a long way.
- Find out their birthday and send a simple card.
- When you identify their interests, keep your eyes open for events in the newspaper that may fall into that area of interest, then cut out the ad or article and mail it to them.
- Is your customer involved with any volunteer organizations? If so, would it kill you to do a little volunteer work alongside your customer?
- Research their interest on the Internet. If you find interesting links, email them to your customer.

One time I had a customer who had an interest in Civil War reenactment. I am not a huge Civil War fan, nor do I really care about people dressing up as the Union and Confederate Armies and pretending they are shooting at each other. You know what? I went to one of these things that my customer was doing, and it was actually pretty cool. Yes, it did kill most of a Sunday, but I was certainly the only rep there, and my customer never forgot it. Trust me, this works and works well. You never know, you may find a new friend or a new interest.

## *Selling to the Big Guy in the Corner Office*

The first time I heard someone use the term "C-level selling," I am almost embarrassed to say what I thought that meant, because none of my customers lived in Florida at the time. I do not care if you are selling widgets or oil part supplies; C-level selling is a term that has taken on more importance this decade, especially in the medical sales arena. First, if you have no idea what I am talking about when I use the term C-level selling, I am referring to the art and science of selling to the people in the executive suite. The "C" stands for "chief," the first letter of the abbreviated title of these executives: CEO, COO, CFO, CIO, CHOO-CHOO (Gotcha!), CNO, and so on. In a hospital, these are the people that have the least involvement in patient care but the most responsibility with regard to how those patients receive their care and the quality of care. Really! You may be asking, how can that possibly be true? I am quite glad you asked.

The products and bedside care provided by the hospital may not be under the direct responsibility of any of these C-level people, but I guarantee that the CEO, CFO, etc., have something to say about how the hospital's money is spent. If the chief of Radiology needs a newer version of an MRI machine and that machine is close to $1 million, it makes sense that the CFO and maybe the CEO need to have or see some kind of justification for that kind of expenditure. We are talking about spending $1 million on one piece of equipment. How can the hospital justify that? Well, what does the facility charge for each MRI scan? What do they charge for an MRA or MRV? How many procedures can they do in a week or a month? How many procedures will be reimbursed by Medicare and private insurance and at what rate? How many scans will have to be completed and reimbursed before the machine breaks even? These are not generally the questions that a radiologist will typically be asked to answer. The radiologist is a physician trained in reading images, not an accountant. The CFO, however, is just the right person for the job of figuring out if the hospital should buy the latest MRI machine or stick with what they have.

So I know you are asking, how does this possibly help me? I have just spent almost a whole book telling you about how to sell your goods and services to the clinical and fiscal buyers, right? I laid out the whole sales scenario for you in Chapter Three. We discussed how Mary was the clinical buyer, and Dave from Apex wanted to sell his product, and how the whole deal went down, but now I am talking about selling to the CFO? Many of you may never get the chance to sell to the C-level, or C-suite customer, but if you can find a way to do it, that will be your quickest way to success.

In the real world of medical sales, this is how it might work using the MRI as the example, but keep in mind, nursing homes and home care agencies have C-level people too. Don from Xepa (that is Apex spelled backward) Corporation sells the world's most advanced MRI machine. This machine can do it all and do it better than any other MRI on the market. Dr. Fitzgerald, the chief of Radiology (once again, not a real person), talks to Don about this new amazing piece of hardware. Dr.

Fitzgerald wants it, and he wants it now! Don knows that hospitals do not fork over $1 million for the purchase of anything without some serious consideration. Don has already sold the doctor on the machine. At this point, Don has one of two choices: First, Don can let Dr. Fitzgerald take the product through the various committees explaining why he needs the machine, and then Dr. Fitzgerald will try to sell the Materials Management Department on why the hospital needs to purchase it, which is always because of the good of the patients, in order to help him do his job better, and to give more accurate tests, bla, bla, bla. Don's second choice would be to take the machine purchase through the proper channels himself by starting at the end. Huh? What does that mean, starting at the end?

If Dr. Fitzgerald was able to manage to take this slick, new MRI machine through the entire purchasing process, which is difficult, a purchase like this would still require sign-off by the CFO or CEO or both. Like I said, hospitals try not to spend millions of dollars without thinking about it, although they do it all the time, but they just do not realize it (that's another story). What if Don started with the CFO? What if Don prepared reports (make sure they have lots of numbers and colors and columns for the CFO—they like that) that showed the projected weekly revenue provided by the new machine compared to the old one? What if he had projections based on the increased reimbursement for a machine this sophisticated? What if he could show the CFO in black and white that the machine would pay for itself in twenty-four months and be producing upwards of $600,000 in annual revenue after two years? If Don could get this information to the CFO directly after getting the go-ahead from the doc, he would be eliminating a myriad of steps normally required to close a deal.

You may be thinking something like, "Okay, I don't sell million-dollar MRI machines, I sell disposable garments or I sell distribution services." The song remains the same. Selling products or services that can benefit the hospital through increased revenue, decreased costs, or streamlined efficiencies can help you sell to the C-suite people. If you can build your case and present it to the C-level buyer, you have a much stronger chance of success. One little warning: This is not for the faint of heart. If you get a C-level meeting, bring in the big guns from your own company, but listen to what they say and what they do. Someday, this will be you presenting to this C-suite buyer.

The trickiest part of selling to the C-level buyer is not the presentation but getting the introduction. For this, you will need the help of your clinical buyer or a friend of your clinical buyer or his or her supervisor, etc. Try to always get "referred" into the C-suite sale. Maybe that referral will come from one of the doctors, which is a great referral source. Even the CEO and CFO care about a physician's opinion. Many doctors today interview with C-level people when they get their job, so there will almost always be some recognition there. If you are dealing with nursing, most hospitals now have a CNO, chief nursing officer. This person is a nurse, and your nurses have the best shot at setting up a meeting with them. Even though they have a clinical background, they are still in the corner offices with the other C-suite buyers. IF you get a meeting with a C-level person, always try to get a referral to the other C-level people. If the CFO sends an email to the CEO to meet with you, he is going to meet with you. Do not be afraid to ask for the referral. Many knowledgeable people have written great books on selling to the C-suite buyer, and I encourage you to read and review them. Reading is knowledge, and knowledge is power.

# Chapter Twelve

## *Presentations: The Good, the Bad, and the "Why Did You Waste Our Time?" (The Third Mystery)*

The sales presentation: This may be the key to success or the nail in the coffin, depending on the rep. Customers in the medical arena will be very honest about the quality of a presentation—maybe not to the person who gave the presentation, but certainly to the people who count and to the competition. Other than listening to the needs of your customer and clearly understanding those needs, there is no part of the sales call or meeting that is as important and impactful to the customer as the actual presentation. Some people are good sales people but not great presenters. Other people are great at the technical part of the presentation, but they cannot close the sale. As a medical sales professional, you want to be good at all facets of the sale, but the sales presentation is probably the part for which you have the most control over, when it comes to how that presentation goes.

If you are the one giving the sales presentation, then the presentation will only be as good as your ability to present, correct? If you are good at presenting your product, the presentation should go well, if you suck at presenting your product, you should find another job. Anyone can be a good presenter as long as they are willing to put in the right amount of time to prepare and practice. As I mentioned during the chapter on committee selling, I always practice my presentations, and I am known for being a good presenter. This was not always the case. When I started in sales, I thought that you kind of memorized a basic schpeel and then answered questions. Wrong! The reasons my presentations are good, I believe, is because of the preparation I put into them. When I know I may be asked to discuss a particular product, I do not just learn the basic features of that product. I learn everything I can about my product, the general product category, the market area of this type of product, the competition, and anything else I think might come up. Why do I do this you may ask? Simple. When I go into a presentation over-prepared, I feel supremely confident in my abilities. I am not worried that a question will come up or a subject area will be broached to which I have no response, because I have taken the time to prepare. I hate to use sports analogies, and most of the women reading this book will agree, but I cannot think of anything better. In baseball, if you want to be a great infielder, you take tons of ground balls. You do not just have someone hit you grounders until you catch one, you catch them over and over until catching a ground ball is second nature. If you have taken enough ground balls, you do not care if they are to your left, your right, one hop, two hop, or at your feet. Once you can do them all, you are prepared to enter the game. That is how I view sales presentations. The presentation will be as good as the work you are willing to put into it.

Great presentations stand out. Unfortunately, bad presentations stand out too. Your presentation is your chance to tell the listener—be it a committee, the CFO, or your main clinical buyer—what you really want them to know about you, your product, and your company. As I stated earlier, your presentation has to be very well prepared and convey the message you want to convey to the buyer or potential buyer. The presentation is the part that takes place after you have asked questions, established needs, etc. If your presentation is bad, your customer will wonder why you wasted their time. Once again, you can go to any book store and find a slew of books on creating and executing a great presentation, but these are some bullet points of what your presentation should convey:

- Introduction – This part will tell the customer, basically, what you are going to tell them, and this is where you may re-state their need.

- Body – This is the exciting part where you talk about the features and benefits of your product, and hopefully you will address the main points that will minimize the questions and objections.

- Summary – Summarizing the key bullet points about how this product or service will help your customer is of the utmost importance. During the summary, you may want to re-state the things you emphasized during the body of the presentation.

- Questions – Address the concerns or unclear points your customer may raise with regards to your product or presentation. If you get a lot of these, you want to tweak your presentation to address more of these potential issues.

• Restate and Clarify – Never just answer the questions and try to close. Always re-state the issue they raised and remind them how your product or service addresses that issue. MAKE SURE YOUR CUSTOMER AGREES THAT THIS ISSUE HAS BEEN ADDRESSED AND SOLVED, OR DO NOT GO TO THE NEXT STEP!

• Get the Lay of the Land – Find out what hoops you will have to jump through in order to get this product in this facility. Does it go to committee, how much product to evaluate, etc.?

• Close – More on this later, but the close is part of your presentation. You want to ask for the order, commitment, evaluation, whatever the next logical step may be. Yes. The close *is* part of the presentation.

# Chapter Thirteen

## *Samples and Evaluations: How to Not Waste Company Money and Your Time (The Fourth, and Most Misunderstood, Mystery)*

I hope you realize how lucky you are that you are learning from my mistakes and that I am humble enough to share these clunkers with you. This area is one near and dear to my heart, and I hope the reader will pay particular attention to this information. Doing it the wrong way can cost the rep and the company money, time, and possibly the whole deal. In many cases, customers, especially clinical customers, have to try a product before they can determine if it works or if it is appropriate for that facility. That makes sense, but like I said before, you can learn, as I did, from doing it the wrong way, or you can simply learn from my mistake I made in 1986.

Earlier I mentioned that I was selling an exciting new product for the treatment of chronic wounds. You may also remember me mentioning how excited I was to sell these new products. When I called on a wound care person, I would talk about the products and give them a certain amount of samples to try on their patients with wounds. This was a set amount of product the company recommended for most customers. In most cases, the customer would use the products right away, and I would follow up in two weeks. For the majority of my customer base, I heard that the products worked great, and they wanted to order them for stock in their facility. One time, however, this scenario did not play out as planned.

In early August, I detailed my products to an ET nurse at a large hospital in the same manner as most of my other customers. This nurse was excited about the products, and I left her with the usual amount of samples. In two weeks, I followed up by phone, and she told me that the products worked great, but they needed to try it on more patients to really gauge the effectiveness of the new products, so I dropped off more samples and scheduled a face-to-face follow up in another two weeks. In two weeks, I met with the clinician, and she glowed about how great the products were working *so far*. So far! I should have known, but I was so excited, I forgot the basics. This nurse needed to try the products on some additional patients, because they needed to follow the patients to the closure of their wounds. Okay, I gave her more FREE product with a plan to call her in two weeks. I called in two weeks, and she was so enthusiastic about how well the products were working, she got me excited, and I could not wait for this large hospital to bring these products into stock. The only problem this nurse had was with one of the doctors who had been in Europe for the past six weeks. This doc was a key guy for ordering wound care products, and this nice doctor wanted to see what all the fuss was about, so he needed to try it on some of *his* patients. I know what you are thinking:

I am just an idiot! Hey, it happens, but at least I learned from it. Begrudgingly, I agreed to bring in TWO MORE WEEKS worth of products for that doctor's patients, with a commitment from the ET nurse that they would be able to place an order at the end of that two-week evaluation.

In the end, this hospital became one of my biggest customers, and that nurse is still a friend today, so it all worked out, but if I would have done it the right way in the beginning, I could have avoided two months worth of evaluation, or I could have at least known that the evaluation, or whatever it took to get the products in, would take two months. The message here is control your evaluations. If, in that initial sales call, you ask your clinical buyer to tell you what the entire process will be to get the product in stock, you will know what you are up against, and you can hold your clinician to that process. Here is a list of questions that will help you bring everything out into the open for determining how many products, how long, etc.:

- "How many patients would you need to try my product on before you decide it is something you want to bring into stock in the facility?"

- "How long would you need to evaluate my product to make a determination on whether or not you want to bring in the product?"

- "Who else would have to trial the product in order to bring it in?"

- "If you like the product, what is the process the product will have to go through in order to get it into the facility?"

- Is there anything else I would need to do after the evaluation in order to get this facility to stock my product?

Restate all these facts with something like, "Mary, if I understand you correctly, you will need to try X number of my products on X number of patients in order to decide if you want to bring my product in for routine use?" "I also understand that you will need to try this for ten days in order to see if the product will work on ten patients, is that correct?" Get all this information up front, and hold your clinician to this information. If your nurse told you that she only needed to see this work well on ten patients, and you brought in enough products to use on ten patients, she should not come back at the end of the evaluation and ask for more product. Also, if all the parameters are established at the beginning, the pressure is off both of you in finding a time to make a decision or close the sale. The very best thing to do, if it is even remotely possible, is to have the facility buy the product they want to trial and give them a money-back guarantee. If the facility does not like the product enough to start stocking it, you will refund the money they spent on the evaluation. Companies do this all the time.

# Chapter Fourteen

## *Closing the Sale*

This brief chapter may describe the hardest thing for our clinicians to master once they move into their first sales position. I guarantee that some of you clinicians in the reading audience who are hearing this now *will* become great "closers," but this is more the exception rather than the rule, or at least that has been my experience. Clinicians, and I will especially cite nurses, usually are natural caregivers. The desire of nurses to make people feel better is what attracted them to the clinical arena in the first place. As a clinician, you may find it easier to master the technical aspects of your products, in-servicing the products, and helping to educate health care staff. These are all important things to do well, but with many clinicians, this is where their work and talent stop once they get on the sales side of the desk.

In an earlier chapter I mentioned people that were maybe better at presenting but could not close the deal, or people who were good closers but they could not present. If you do not do a good job presenting, you may not get a chance to close the sale, but if you get the chance, you have to act. All the great information in the world does not amount to a hill of beans if you cannot get the customer to place the order. When a rep tells me what a great presentation they had at XYZ hospital, I am happy for them, but I see it as meaningless until we close the deal. Closing the sale is what pays our checks! Most companies do not pay medical sales people to do presentations, in-service, or educate the customer. The companies pay sales people TO SELL! You cannot sell if you cannot close. I have seen an awesome array of books on closing the sale, and I recommend you get at least one of these and read it cover to cover. In addition to teaching better ways to close a sale, most of these books help you feel more comfortable with asking for the order, because they describe closing as a natural part of the sales cycle. Sometimes this is the biggest info, the "comfort" sales people have to develop in order to ask for the business. I can tell you that if you let yourself feel like you are asking your mom or dad for a loan, you will not be successful. Think about the close as the next natural part in the sales cycle.

People, even clinicians in health care, expect a salesperson to ask for the order. I have been fortunate to have clinicians and purchasing people share stories with me about fellow reps who could not, or did not, ask for the order. I can vividly remember one nurse telling me, when speaking of a quasi-competitive rep, a story that ended with, "I waited and waited for her to ask for the order and she didn't. I sure as heck was not going to give her my business if she didn't ask me for it!" This was coming from an advanced practice nurse in a major hospital. See. Closing is important.

Unfortunately, when many people think of "closing"

they think of our aforementioned used car salesman. Closing does not have to be this way and should never be this way in health care. Health care professionals expect to be treated like professionals, and the best way to close a professional is to let them close themselves. Huh? Those of you who are reading this that come from a successful sales background know exactly what I mean, but for our clinical folks without the benefit of your experience, I will describe exactly what I mean.

Remember, I talked about asking questions to identify the true need or pain of your customer? If you ask the right questions, you can direct the sale to the natural close without even appearing to "close the deal." If you set up the evaluation the right way in the first place, by identifying the process, the length of evaluation, number of patients, amount of product, committee process, etc., you are 90 percent of the way there. Before you left that nurse's office when you set up the evaluation, hopefully you asked her a question similar to this: "If you go through the evaluation and committees, and you and your facility like the products and they are approved, would there be any reason why you would not bring in my product?" As you are hand-holding the evaluation and taking your product through the committees, you will already get a good idea if your clinician likes your product. If your product works like you say it does, and it is priced appropriately, your main contact will already want to bring it in. In effect, he or she will already be closed before the committee meets. If this is the case, you only have to reaffirm with the clinician that the product will be brought in once the committee approves it. Make sure you do this before the committee meets. You could say something like, "Once the committee approves the product, how many units should I have Fred in Purchasing order?" Or, you can be very assumptive and ask if we can set up in-servicing for the floors in anticipation of the committee approving the products. Hmm? A little bold, but not out of line.

The bottom line is this: You do not want any surprises to pop up anywhere along the way. If you have done your job from the beginning, including asking questions about the number of patients needed to use the product, the amount of product, who else had to be involved, supplying the proper amount of product and in-servicing to complete the evaluation, etc., there should be no surprises. If you have done your job, closing or assuming the close should be a natural progression. Unfortunately, some reps are not comfortable asking for the business, as I said previously. Do not become one of those people. I hate to see a rep who is great at detailing the product but cannot close the deal. I would rather have a mediocre product person who can bring home the bacon any day of the week.

# Chapter Fifteen

## *Service, Service, Service*

Service? *I can hear your questions/comments already.* What if I am not in a service business? What if I sell a product? This chapter is not in the book to cover service businesses to health care, this chapter is going to tell you how important *service* is to successful sales. The junkyard of failed companies is littered with the debris of poor service. Let me ask you a question. If you buy a car from a dealer that is not a one-of-a-kind car dealer (think Ford), and you need service or routine maintenance, are you likely to take the car to that dealer, at least initially? Probably so, provided the dealership is reasonably convenient in location and you felt comfortable with their Service Department when you visited the dealership. If you take it in for service and the car is not fixed correctly, or the staff in the Service Department has a bad attitude, or maybe you have to wait longer than normal for the work to be done, then how likely are you to want to continue using the dealership's Service Department? Probably not that likely, but you may be stuck, because maybe this is the only Ford dealer close to your house, and you do not want to drive a long distance out of your way for service. Yet, maybe the service is bad enough to make you drive out of your way just to avoid that dealership.

Sales, and medical sales, can be the same way. Really! My experience shows me that clinicians hate the strafing run—type sales people in my business. What do I mean by strafing run? These are the reps that come in strong, make the presentation, get the sale, and then move on to the next conquest, rarely to be heard from or seen again. I know that we all want to get as much business as possible, and I am all for closing a deal and looking for more new business, but you have to service your current business in order to keep it.

So, how do you service existing customers? I am hoping that you previously received substantial training for this area in your last sales job, but just in case, let me give you a few bullet points that warrant your attention. My dictionary lists twenty-six different entries for the word *service*.[10] The very first entry refers to *service* as "an act of helpful activity." How simple can one get? You will provide service for your customers by anything you do that can be considered "an act of helpful activity." Why don't I list some "helpful activities" for you.

- Keep in touch with your customer to make sure things are going well with your products/service.

- Return phone calls and emails just as quickly as you did when you were trying to get their business.

- Provide regular in-service education for the account. (The frequency is something that will be determined by you and your customer.)

• Continue to show an interest in your customer's life.

• Continue to entertain and build relationships where possible.

• Keep your customer informed of changes in the market, your product, and the competition.

• Make sure you continue to schedule regular appointments with your key contacts at that account. (The appointments may not be as frequent, but if you are smart, you will make them as regular as reasonable.)

A great book on this subject is *The Successful Manager's Guide to Selling Through Proactive Customer Service,* by Lee Van Vechten. This book discusses the importance of proactively providing strong customer service and response in order to boost sales and keep customers as buyers in the long term. I am sure you can find many good books on the subject as well, but the point is to keep servicing your customer to make them continue to feel that they matter, and also that their business and opinions still matter to you and your company.

More than five trillion times (Okay, maybe not five trillion), customers have told me that they never see "so and so" since the customer agreed to buy their product. Eventually, that person they never see is going to lose that business to the first sales rep that walks in the door with a decent alternative. This is especially true when it comes to proprietary products. In the beginning, the company and the rep think that they can slack off on service after the sale because the competition for that product or service is limited. Please believe me when I tell you that no matter what your company produces or sells, you WILL have competition sooner or later.

When I started in health care sales, we had a product that was truly unique, and this was the type of product that showed tremendous results with almost no competition. Our sales during the first few years grew exponentially. Our company was doing well, the market loved our company, and the sales reps were making good money. Unfortunately, our company was publicly traded, subject to the whims of a board of directors. At one point, the people that built the greatness of our company were forced out, and they were replaced by people LARGELY FROM THE PHARMACEUTICAL INDUSTRY, who possessed not a clue as to what we did, how we did it, why we did it, and how to keep doing it. Within twelve months the company had raised the price on our unique product line three times, alienated our core clinician to go after a different primary call point and informed the first GPO in the country that they could go fly a kite if they thought they could negotiate special volume pricing. This sort of combination should be described in sales manuals as what *NOT* to do.

Meanwhile, many of our competitors were testing similar-type products to bring to market. The first place they all went was to those core clinicians our company did such a nice job alienating. The clinical buyers tried the new competitive products, and if they were being honest, they would probably say that the new products did not work quite as well as my product, but they worked well enough to warrant a look-see, especially since all of the competitors were 20–40 percent cheaper than my product. This was a hard thing to overcome, and we absolutely lost some business. That exponential sales growth I mentioned—well, that stopped. As a matter of fact, almost all of our sales growth stopped. Fortunately, the board of directors was not completely stupid, and they recognized a need for change. The board brought one of the key early successful people from the old days back, lowered prices, and apologized to the multiple GPOs they alienated. In this situation, unfortunately, the damage was pretty much done, and little could be done to save that division of the company. Our division was eventually sold to a larger company that recognizes the benefit of strong customer service, and since that sale, our division has grown much faster than our competitors. This entire saga could have been avoided if the company recognized the importance of service. Never forget: service sells, a lack of service destroys. The old saying in the sales world is "Take care of your customer, or someone else will." How true!

## *In-Servicing*

Another aspect to service is the training and education of the facility staff. The reader may be lucky enough to sell the best product ever introduced to the health care market, but if the staff does not know what that great product does and how to use it, they are less likely to use it and less likely to feel positive toward its use. After you have closed the sale and walked your product through the process of getting your box on the hospital shelf, the hard work may be just beginning. Huh? That's right. You have to train and educate the staff on this new gizmo of yours.

During the evaluation process, hopefully you already were able to in-service your products to begin the evaluation, but even if you did, the importance of in-service education cannot be emphasized enough. In-service is not only the time that you train and educate, but also the time when you get to sell more product. The more the staff likes and is comfortable with your product, the more they will use it. The more they use it, the more they will need it, thus they buy more product. I always tell the reps that the in-service is actually a great sale opportunity. If you have the business at a hospital, and your sales are flat, set up additional in-services facility-wide, and I guarantee the usage of your products will increase.

Your customer, and probably the Education Department, will help set up the in-services in that facility, but be prepared to work hard. Most facilities want all three shifts, along with weekend staff, trained on the products. This means that you will have to be there for those in-services. Yes, you will have to give up some weekend, evening, and sleepy time in order to get your product fully integrated in the facility, but it is well worth it. Just a side note: The more help you can provide in making the scheduling, implementation, and execution of facility-wide in-services for your customer, the better reputation you will carry for the next customer. In-servicing is hard work, and the more help you can lend, the easier it makes things for your customer. Here are a few bullet points of things that are important for successful in-servicing, but always follow your customer's and your company's guidelines:

- Try to schedule roving in-services. This is where you go from floor to floor educating people when they are not busy. The alternative is setting in a room and waiting for people to come to you. (This rarely produces a good turnout.)

- Try to never agree to in-services where you are stuck in a room waiting for people. (Do you get my emphasis here?)

- If allowed, bring snack-size candy bars to hand out for staff that attends your in-service, except for mornings, then nothing says sales like donuts and pastries.

- Keep it simple, quick, and basic.

- Make sure everyone is comfortable with your product. If they were already using the product during the evaluation, make sure you answer any questions that came up.

- Try to provide your customer with education posters and flyers for use in educating the staff. These tools provide education and service after you leave.

- Keep your primary customer apprised of your progress throughout the schedule.

- Encourage all who attend your in-service to use your product. Your product would not be in the hospital if it did not have a benefit for the patients and/or staff.

Try to re-in-service your facilities every six months. I know that this sounds like a ton of work, but it will be worth the effort. Staff comes and goes. Doctors come and go, and people need some re-education from time to time. Plus, this ongoing training helps keep your sales strong.

# Chapter Sixteen

## *Reimbursement: Who Pays for All That Stuff? (The Fifth Mystery)*

Who *does* pay for all that equipment, all those supplies, and all that medicine? This is a good and fair question, and you may be surprised at the answer. We do!!! That's right . . . you and I. One way or another, we all end up paying for the majority of costs associated with patient care. Most Americans do not understand, but the vast majority of people in our country do not pay the direct costs associated with their care. For example, if you have gall-bladder surgery in a hospital, and you spend at least one night as an in-patient in the hospital, you will have a bill of at least $11,000 between the surgeon's fee, anesthesiology, operating room, recovery, and finally, the time you spend on a Med/Surg floor. Here is the question: Did you pay that $11,000 out of your own pocket? Probably not. So who *did* pay the bill?

### *Private Health Insurance*

Most Americans have health insurance (despite what the media wants you to believe) either through a private insurance health plan, employer-sponsored health plan, Medicare, Bureau of Indian Affairs, or Medicaid. Even with private insurance, you do not pay the $11,000. Your insurance company pays that, and you or your employer pay a monthly, quarterly, or annual fee to cover your "premiums." In the preceding example, let us say you were paying $250 out of pocket each and every month to cover your insurance premiums. The insurance company is taking that $250 and betting that if you get sick, your illness will cost them less than your $250 per month in money they pay to the hospital, surgeon, etc. On a one-on-one basis, this would be pretty risky for the insurance company, so they spread the risk over a larger number of people. Your company might have two thousand employees, and the insurance company is betting that if all two thousand employees pay $250 per month for health insurance ($6 million annually), the insurance company will incur costs that are less than $6 million that year. Actually, they would like that cost to be less than $4.8 million in order to have a decent return on investment and to hedge their bets for next year. In case you haven't noticed, large insurance companies do not tend to go out of business very often.

Private insurance pays for many of the costs associated with patient care, but private insurance is not the largest payer of health care costs. The largest payer for health care expenses incurred in the hospital, nursing home, home care, or hospice care is a coalition formed by citizens like you and I called the government. Okay, we didn't necessarily form the coalition, but you know what I mean. We pay the costs of these state and federal health care programs through our generous donation to the state and federal government every year in the form of taxes. Yes, that is correct, anyone who works supports the federal and state health care system through the payment of state and federal taxes used to float the cost of health care. Even if you are fortunate enough to live in a state without a state income tax, part of your federal Medicare payment supports your own state health care program, usually called Medicaid. In California it is MediCal, and in Illinois it is Illinois Public Aid (IPA), but whatever the entity is called, you and I pay for it one way or another.

(Medicare and Medicaid Disclaimer: I am in no way, shape, or form a Medicare billing expert. All of the information contained herein is from my experience with real customer situations and all information is gathered from our government's own Medicare website, www.medicare.gov, or www.cms.gov. If you work with a third-party billing company or sell products that are reimbursed through a third party, make sure you follow all proper company procedures and protocols. Unless you become a reimbursement expert someday, you are NOT a reimbursement expert, nor should you ever give reimbursement advice to your customers, or anyone else.)

## *Medicare*

The Medicare program was established by Congress in 1969 as a way to fill the gap for retired citizens who did not have retirement health care coverage from an employer, and this program is administered by the Centers for Medicare/Medicaid Services, also know as CMS (formerly called HCFA). In today's world, most people who turn eligible age for Medicare coverage use it. Under Medicare Part A, all U.S. citizens over age sixty-five are eligible for hospital coverage and limited nursing home, home care, and hospice coverage.[11] The time is limited during which Medicare Part A will pay, but everyone has that coverage without paying a penny out of their pocket for a given period of time. Medicare Part B is a supplemental policy available to anyone who qualifies for Medicare, but they must pay a monthly premium ($87.50 per month in 2006, but in 2007 this became a sliding rate based on income) in order to have supplemental coverage once Part A coverage ends. Medicare Part B primarily covers certain physician and diagnostic services, enteral feeding, catheter supplies, ostomy supplies, durable medical equipment, and surgical dressings. More on this later.

If a patient is in the hospital for less than 100 days, or the patient goes to a nursing home within that 100-day period, the patient is covered under Medicare Part A for almost all costs of care in the hospital, nursing home, or home health care. Once the 100 days has expired in the hospital or nursing home, the patient switches to either private insurance, private pay (cash out of pocket), or Medicaid in most situations. Medicaid not only kicks in to cover patients who no longer have Part A coverage, in most states it covers lower income, younger patients who do not have health insurance of any kind. In most states, Medicaid is a stressed system which does not routinely reimburse well. You may find this information most helpful if you are selling to the nursing home or home care market.

Speaking of home health care, few patients have coverage for in-home care through private insurance, although more and more people are investing in this type of insurance within their "long-term care" insurance plan. Home health care, in most instances, takes place as we discussed . . . in a patient's home. The care at home is primarily reimbursed through Medicare Part A, and the costs of this care are reimbursed based on the acuity, or sickness of the patient. Since 2002, home health care agencies are reimbursed on what is called an "episode of care," which is usually a sixty-day period of time for which the agency gets paid to care for the patient. The home care agency receives a flat dollar amount for the entire medical aspect of the care for that particular patient for sixty days. How much do they get, you ask? It depends on how sick they are, and how much therapy they need. Like the MDS-created Resource Utilization Group (RUG) categories for nursing homes I will cover later, home care agencies use Oasis, an evaluation software, and based on what Oasis calculates the agency will get this flat amount for all costs associated with the patient's care. This amount averages roughly $2,100 per sixty days, which for some patients is more than enough, and for other patients not even close to enough. Because of this type of reimbursement, the modern, twenty-first-century home care agency must provide quality care and be fiscally responsible. If the care is not quality, and the patient does not improve, without a very good reason, the home care agency will stop being paid. I will discuss more in the next paragraph, but the new system to be implemented in 2008 will be similar, but there will be a separate dollar figure associated with the costs of supplies, which could increase their overall reimbursement. However, CMS says that the change will be "revenue neutral."

CMS released their final version of the PPS payment rules for home health agencies on August 22, 2007. Most of the stuff we just covered under the Home Health reimbursement area did not change. However, the amounts and emphasis is changing for 2008. According to this latest report, these are the major changes:[12]

• The Base Rate, or average, ($2,100) will be reduced by 2.7 percent . . . ouch! The government feels like the home care agencies are making too much money off of too many patients.

• The way therapy is reimbursed will change. In the past, the government provided a significantly higher payment if a patient needed more than 20 therapy visits, and guess what? A ton of patients seem to need exactly 21 therapy visits. Huh! Isn't that a strange coincidence? Now the reimbursement will be reduced overall and consist of three levels: 0–13 visits, 14–19 visits, and 20 or more visits. This area will be much more closely scrutinized for payment.

• Data on specific Quality Outcomes will have to be included with each final submission of the UB-92 form. (The UB-92 is the form the agency has to submit in order to get the rest of their fee). Agencies will have to *prove* that they are making patients better in addition to actually making them better.

• That small part of money that was given for medical supplies is being taken away completely, to the tune of almost $50. In the place of that flat rate, the agency will have to complete an additional set of questions on the Oasis regarding supply needs. The agency will then see a separate reimbursement of between $14 and $500 for that patient's episode. If tracked properly, the medical supply reimbursement could actually be a source of revenue for the agency, and if you sell for a company that offers cost-savings products or services, so could you.

All of this may sound bleak, but if an agency has been proficient at tracking supply costs and has provided high quality care, they could actually benefit from the changes in reimbursement.

## *Medicare Part B and Part D*

When we think of Medicare, and when health care providers and Congress discuss Medicare, they are normally referring to Part A and things associated with Part A care and payment. The aforementioned Medicare Part B is a totally different area of reimbursement. As I stated earlier, Medicare Part B is a supplemental policy to which a patient or patient's family must pay a premium in order to have this type of coverage. The largest difference with Medicare Part B is the provider. Medicare Part B services are provided by a third-party billing and supplier company, and sometimes by the nursing home, *if* they have their own Durable Medical Equipment (DME) provider number. Under the old system, third-party billing companies provided the enteral products, including pumps, tubing sets, and even the liquid nutrition, as well as ostomy supplies, wound care surgical dressings, durable medical equipment, catheter supplies, etc., to the nursing home and then billed the regional insurance carrier licensed by the government. This same regional carrier, which is kind of like a regional insurance company, now called Statistical Analysis Durable Medical Regional Carrier (SADMRC), in turn would process the claim and pay the third-party provider directly. Part B providers can make a profit under Medicare Part B, consequently, intelligent nursing home owners obtained their own provider numbers and bought and billed the products themselves, and many continue to do this today. (Note: In 2006, CMS awarded contracts to four regional carriers specifically for the processing of claims related to durable medical equipment, custom braces, prosthetic devices, and respiratory equipment.) This arrangement provided a new revenue stream, but it also required additional staff to bill Medicare Part B properly, in order to make it work. Many companies across the United States are in the sole business of providing medical supplies under Medicare Part B for nursing home patients or patients at home not covered under home care.

Under Medicare Part B, the federal government covers 80 percent of the cost of the goods or service, then the provider has to bill the secondary insurance carrier, state Medicaid, patient, or family to collect the remaining 20 percent. In many instances, the 20 percent "co-pay" goes uncollected, because many families and patients do not have the capacity to pay. If that patient's state does not cover the co-pay, and the patient does not carry a supplemental Part B policy, and the patient does not have the money to pay, the provider has to eat that 20 percent. You may ask, why would they ever eat the 20 percent? Sometimes the reimbursement amount for an item may be 60 percent over the cost of that item. If the provider does not get the remaining 20 percent co-pay, they still made a nice little profit. If everyone did that, however, Medicare would say, "Okay, you guys are content with the 80 percent we are giving you, so we are going to reduce their overall reimbursement, because we are obviously overpaying." Medicare REQUIRES the billing company, generically referred to as "Third Party Billing" companies, to bill the co-pay until they can document that the co-pay is not collectable. You may be asking, what happens in the real world? Pay attention:

> ABLE DME Provider Company buys a surgical dressing for $10 (their cost). After all proper documentation and paperwork and all appropriate signatures (Face Sheet, Copy of Card, 1500 Form, CMN, etc.), ABLE DME provides a two-week supply of this daily dressing to the patient at no charge to the patient right now. According to the Fee Schedule and HCPCS, the "Allowable" for that dressing is $20. (See next page for HCPCS schedule).
>
> The Provider, ABLE DME, electronically bills the SADMRC $280 ($20 × 14 days) and supplies all the necessary paperwork to Medicare. At the same time, ABLE DME sends a bill for $56 to the patient's home. This is called the co-pay, or the co-pay could have gone to an insurance carrier or state agency where the patient lives that happens to cover Part B gaps.
>
> In fifteen days, Medicare sends a check to ABLE for $224. (80 percent of $280). Right now, ABLE sees a gross profit of $84. This is not much, but it is still nearly 40 percent. If ABLE could get the co-pay, they would see an additional $56 added to their bottom line.

The government requires the DME provider to continue to attempt to collect that co-pay until it is no longer financially practical. There is not a hard and fast rule on this, but according to providers I interviewed, providers are expected to bill at least three times for the co-pay, but leave that issue up to the experts.

Let's look at the financials of providing products under Medicare Part B:

| | |
|---|---|
| *Dressing Cost* | *$140 (Two-week supply)* |
| *Medicare Submission* | *$280 (Two-week supply)* |
| *Medicare Payment* | *$224 (80 percent of the Allowable)* |
| *Patient Co-pay* | *$56 (Still to be collected)* |
| *Profit w/o Co-pay* | *$84 (38 percent gross profit) ($6.00/day)* |

The provider was able to make $84 for two weeks, or $180 per month on just this one patient. If this was a surgical dressing and the patient had two wounds and this provider was providing supplies for all twenty patients at the nursing home with similar wounds, they would be realizing a gross profit of $7,200 without the co-pay. Can you see why the DME provider is not going to cry if they do not get the co-pay? Some payments are not as lucrative and the provider cannot afford to provide the product without the co-pay, but some are even more profitable. The point here is to get to know the companies in your area, if any, that supply your products as a third-party billing company.

Under this scenario, the company made a decent profit, but it is not always that way. In some situations, the reimbursement (allowable) does not even cover the cost of the product or service, or it provides room for much smaller profits. In addition, companies often have to jump through so many hoops that it almost does not make it worthwhile to even be in the business of proving DME supplies. A colleague told me about one large third-party supplier who had over $300,000 in Part B claims waiting for review. This means that the company has to supply tons of documentation in order to get paid, and it also means that the payments have been delayed for an extended period of time. The good companies that stay in business and make money are usually the ones who are following all the rules and providing a quality service for Medicare beneficiaries.

| *Code* | *Description* | *Action* | *Replacement Code* |
|---|---|---|---|
| A4213 | Syring, sterile, 20 CC or greater | Add | |
| A4215 | Neede, sterile, any size, each | Add | |
| A4348 | Male external catheter with integral collection | Delete | |
| A4359 | Compartment, extended wear, each (eg, 2 per month) | Delete | |
| A4244 | Urinary suspensory without leg bag | Add | |
| A4245 | Alcohol or peroxide, per pint | Add | |
| A4246 | Alcohol wipes, per box | Add | |
| A4247 | Betadine or phisohex solution, per pint | Add | |
| A4461 | Betadine or iodine swabs/wipes, per box | Add | Replaces code: A4462 |
| A4462 | Abdominal dressing holder, each | Delete | Replacement code: A4461 & A4463 |
| A4463 | Surgical dressing holder, non-reusable, each | Add | Replaces code: A4462 |
| A4932 | Rectal thermometer, reusable, any type | Add | |
| A6412 | Eye patch, occlusive, each | Add[13] | |

## Medicare Part D

Medicare Part D is the newest entry into the Medicare reimbursement system. Part D was established in 2005 and implemented in 2006 to provide coverage for prescription medications to Medicare-eligible patients. Like Part B, the patients pay a small premium and a certain amount of out-of-pocket expense. If a patient has little need for prescription drugs, which is rare, Part D is not a tremendous financial help. The patient who, on the other hand, runs through a ton of expensive medication faces a maximum out-of-pocket expense per year, which varies by plan. They also will incur a cost of up to a $265 deductible, plus the cost of the premiums. For some of our seniors, the Part D program has made the difference for them in being able to buy food instead of spending all their money on medications. Medicare Part D will not have much of an effect on what you are selling, unless you are one of those silly pharmaceutical reps that are still reading this book.

## The Good Old Days of Long-Term Care

Medicare reimbursement is not going to be at the forefront of your customer's mind in the acute care arena, but in long-term care, reimbursement can be key. In order to truly appreciate the system in which nursing homes and home care agencies find themselves today, you have to understand the previous system for both classifications. Both systems are explained with broad strokes here, but this review is only meant as a brief overview of the previous Medicare reimbursement system.

## *Nursing Homes and Medicare Part A*

Prior to 1996 and a program called the Perspective Payment System (PPS), the nursing homes were reimbursed based on what they spent. The more you spent, the more you made. Here is an example of how the old, pre-PPS system worked: The nursing home was able to add their cost of overhead, such as heat, electrical, improvements, repairs, staffing overages, etc., into their cost of doing business. If the nursing home paid $1,000 for a product or service, they would bill Medicare, say $1,500, based on their overhead costs. The government would reimburse, roughly $900–$1,100 to the nursing home, and the difference between what they billed, and what their actual cost would be (in this case $400–$600) was placed on their cost report, which made their average daily rate increase for the next reporting period or the next year increase. The more you spent, the more you made. Unfortunately, this was not a system that could sustain an aging population and an ever-expanding Medicare base of patients. The government developed a PPS for reimbursing nursing homes for their costs as a way to save the program.

This new program was not even really based on their costs but on the acuity or condition/complications of the individual patients. The federal government established unique categories for patients based on the projected level of care that would be required during their stay in a Medicare Part A bed. Patients have many different conditions and complications, but the government decided that all patients could be grouped into one of forty-four different classifications, called Resource Utilization Groups (RUGs). But how did a nursing home find a RUG category for a patient? Good question. Each individual nursing home had to complete a computer form called a Minimum Data Set (MDS), and once this form was completed on an individual patient, the computer would spit out a RUG category for that patient. The nursing home would then receive a daily rate based on that patient's RUG category. This daily rate was intended to cover ALL the care and costs associated with that patient. This "facility-specific" rate was phased in over a period of four years, so that in today's world all nursing homes are reimbursed under Medicare Part A at a set rate based on the RUG category of that patient. This type of system incentivizes a nursing home operator to spend wisely to provide the highest quality care at the most economical price. Anyone can provide cheap care, but the quality of the care will suffer, and the nursing home could lose their Medicare certification, which is NOT a good thing.

The original change in payments was phased in over four years. The facilities cost and payment history was calculated before the implementation of the PPS program, and this resulted in a facility-specific rate. The facility-specific rate referred to the daily rate that facility should be paid for a patient with that RUG category diagnosis, based on how that nursing home spent in the past, before PPS. In the beginning, three fourths of the amount paid was the facility-specific rate and one fourth was the Medicare rate. The next year it was 50–50, and so on, until all facilities were operating 100 percent under the government rates by year four of the program.

## *Home Health Care Agencies and Medicare Part A*

As we discussed earlier, home health care agencies were reimbursed on a similar cost-plus formula in the old days, and home care agencies multiplied like bunny rabbits. The government (whether they admit it or not) knew that some of the agencies were not delivering the kind of care the cost warranted, and it is rumored that they established a goal of closing half of the home care agencies in the United States. How does one accomplish this? Easy . . . stop paying—not really stop paying, but pay so little that the operator cannot afford to stay in business. This was called the Interim Payment System (IPS). Interim meaning we (the federal government) don't know what we are going to pay and how we are going to do it, so until we decide, we are going to pay you X amount of dollars, and if you make it you make it, if you don't, you don't. And that is the way it was. In 2004, the government started a type of PPS program for the nation's home care agencies, which was loosely based on the implementation of the nursing home program.

The new Prospective Payment System for home care agencies with Medicare patients is very similar to the nursing home system. The only real difference is the "episode of care" tag and the name of the software used to determine the level of reimbursement for that episode of care. As I stated earlier, this evaluation software for home health care is called Oasis. In today's competitive market, the home care agencies, like the nursing homes, are rewarded financially if they spend wisely and provide high-quality care to the Medicare Part A patient.

Previous to the IPS system, home health agencies were reimbursed in a manner very similar to nursing homes under Medicare Part A. The cost of overhead was figured into the amount billed for the patients care and products, so once again, the more a home health company spent, the more they were reimbursed. In the old days, smart home health agencies re-submitted their cost reports on a quarterly basis, instead of annually, so that they could get an increase in their reimbursement rates on a more regular basis.

## *Medicaid*

This will be a simple and short paragraph. Medicaid, while partially funded by the federal government, relies heavily on state funding to care for patients who have no insurance and do not qualify for Medicare. Traditionally, Medicaid reimbursement rates are lower than Medicare, but Medicaid does pay for almost all of a patient's care if they are unable to pay. Furthermore, most state Medicaid programs cover all prescription medications, but this differs from state to state, and the Part D component is changing for Medicaid recipients in a way which may help the survival of state Medicaid programs. If the patients have Medicare Part D, their drugs will be reimbursed under that plan, and the state Medicaid agency will not have to dip into their state budgets in order to pay for prescription drugs. The Medicaid program is administered differently in every state, but this program *is* 90 percent administered by each individual state government through a state-run agency. With a system like this, each state gets to determine what they do or do not cover. For more information, please visit your state's website or go to www.cms.hhs.gov/MedicaidGenInfo and click on the link "Medicaid Facts and Figures."

## *Bureau of Indian Affairs (BIA)*

Everyone has heard of Medicare and Medicaid, but I am willing to bet that few of the readers have heard of the Bureau of Indian Affairs, or specifically, Indian Health Services (IHS). This is a separate area of government-funded health care that applies only to health care provided to and on Indian Reservations. The benefits provided under BIA are quite liberal, and the patient care on an Indian Reservation is extremely comprehensive. Health of the Native American Indians is poor compared to the general population, especially when it comes to the number of diabetics, rate of venous and arterial disease, alcoholism, and a litany of other health issues, but fortunately for Native Americans, the coverage for health care that is needed is more than adequate. BIA and IHS basically cover almost every type of care and product needed. If you are lucky enough to have your products covered under BIA, you should be calling on the Indian Reservation hospitals and clinics, because it is nearly a slam-dunk. You can find more information regarding this program at www.ihs.gov.

## *Dark Days of the DRGs*

You may be asking yourself, why did all this stuff change? Why does Medicare pay so much less than they used to, and why does it appear to be so complex? The simple answer to that question is the Diagnostic Related Group

(DRG). In the late 1970s, Medicare could see the escalation of costs throughout the health care continuum, and fiscal data clearly showed that hospitals were the main beneficiary of that current payment system. Medicare, with the help of consultants ranging from the field of insurance actuaries, physicians, hospital purchasers, and even accountants, determined that costs could be calculated with some accuracy based on the procedure performed for the patient, and the follow-up care needed.

Beginning in 1983, Medicare created the DRG program, which was quickly adopted by private insurance. This new program paid a very narrow range for the care of a patient, based on the procedure, instead of paying the hospital whatever they wanted to charge, and this flat rate had to cover all costs associated with that procedure or diagnosis.

For example:

| | |
|---|---|
| *Gall-bladder surgery* | *$14,300* |
| *Hip replacement* | *$24,390* |
| *ER visit to set a broken leg* | *$9,800* |

A patient comes into the ER for a broken tibia. The DRG would provide payment to the hospital at a flat amount ($9,800 in our example). If the hospital fixed the leg quickly and properly and sent the patient on their way, they might make money. On the other hand, if the hospital had problems and the patient had to be admitted or go to surgery, the hospital could lose money without an additional diagnosis.

This new system forced the health care providers to be more cost-conscious, and therefore, they had to provide high quality care at a fair price. The DRG system, from a cost-containment perspective, was a huge success. For the hospitals, medical product manufacturers, and distributors, this was not a boom. In fact, it was a large potential bust. For hospitals, the "spend more—make more" philosophy of the past was dead forever.

The DRG theory worked so well from the government and insurance companies' standpoint that this became the broad model for the PPS program for nursing homes and home care. At the time of this writing, the programs are about to undergo an even more dramatic change called, Pay for Performance (P4P). Not only would hospitals be restricted by DRG rates, effective in 2008 hospitals will be penalized for poor care and rewarded for better care, as determined by a set of government guidelines that include things like reduced surgical infections, reduced blood stream infections, reduced return visits for patients, and even reduction in ventilator-acquired pneumonia. Currently, the government has 538 DRG categories, but effective in late 2008, we will see a total of 745 "Medicare-Severity" DRGs. The new DRGs will see three tiers of payment per diagnosis, based on levels of patient acuity or additional details of co morbidity and complications. Without going into too much detail, the new payments will change and incentivise quality care by the hospitals. If you would like more information on this DRG change and updates, there is a great overview at www.healthleadersmedia.com.[15]

In order to sell, it helps to know who pays, how they pay, and how much they pay.

# Chapter Seventeen

## *Staying Out of Trouble: On a Clear Day You Can See Leavenworth*

Why do I include this chapter in a book on selling to health care? By the end of this chapter, that is a question you will be able to answer yourself. The vast majority of health care for sick patients is provided for patients who are older than the average population. With this in mind, it should be obvious to you that much of the health care we see is affected by Medicare and Medicaid. As we discussed in the previous chapter, Medicare and Medicaid are federally funded programs. Any time our government is involved in something, they are trying to make the program better, and sometimes they have to take on the job of finding people who are doing things below-board. Unfortunately, in the earlier days of Medicare, the government experienced a large amount of fraud and abuse. Health care providers were billing for services they did not provide or were providing services that were different than the service being billed. Confusion over reimbursement guidelines did not help, and unlike today's system where only four regional insurance carriers set the rules and process claims, there were twenty-three carriers making rules and processing claims, and every carrier had their own guidelines.

Much of the "fraud and abuse" the Medicare program experienced was actually due to confusion and conflicting guidelines provided by the carriers. In 1993, Medicare went to the four regional carrier programs, and the guidelines, rules, and reimbursement rates became fairly consistent from region to region. In the new system, all carriers have the same basic reimbursement rules, but the rate they pay differs along a "range" provided by the federal Medicare mothership known as the Center for Medicare / Medicaid Services (CMS), formerly known as the Health Care Financing Administration (HCFA). When the regional carrier system was put in place, incidences of "fraud and abuse" dropped significantly. It also was helped by aggressive prosecution of providers suspected of fraud and abuse. In the current system, much of the real fraud and abuse that occurs is truly people trying to scam the system. This aggressive federal prosecution did not go unnoticed by other providers, and the good providers that stayed in business followed the guidelines much more closely.

Despite these changes toward positive reform, the federal government has realized that there is money to be gained from finding and even creating fraud and abuse from large companies. If a large medical or pharmaceutical company is accused of wrongdoing related to health care, the argument is made that all health care dollars eventually get back to having a portion paid from Medicare or Medicaid. If a company would be found guilty of defrauding the government through Medicare billing or whatever, they could lose the ability of their products to be reimbursed by the government. In other words, they would be prohibited from participation in the federally funded health care programs. Unfortunately, hospitals and nursing homes use Medicare, so a ruling like this would put the company out of business, regardless of their size.

In the early part of the twenty-first century, our all-seeing, all-knowing, wise U.S. Congress passed legislation letting the law enforcement agencies who received money as the result of a corporate fine or penalty from a plea arrangement or conviction keep a large chunk of that money to pay for more enforcement. So in essence, if the FBI helps convict or get a guilty plea from a company who pays a fine or settlement of $100 million, a significant piece of that $100 million goes back into that particular agency's budget. Some thought leaders theorize that this seems like it causes the federal law enforcement agencies to do whatever they can to get money from companies. Once the federal government gets on a person or company, people find it very difficult to get out from under their pressure. According to 2005 figures, less than 10 percent of federal health care indictments result in trials. The vast majority are settled before trial or dropped.

In 2003, Abbot Laboratories settled a case brought by the government against their division that manufactures and sells enteral feeding products for over half a billion dollars. According to the company, Abbot felt strongly that they had not knowingly committed any crime, but it was not worth it for them to run the risk of losing the case and being barred from participation in Medicare reimbursement. As was stated earlier, Abbot makes enteral feeding supplies, and almost all the patients in a nursing home with a feeding tube are being paid for by Medicare Part B. Subsequent similar cases were settled or dismissed. This is just one example, but if the reader is involved in selling products or services in the health care market, the reader is working under the guidelines of the federal government.

In 2003, the Department of Health and Human Services (HHS), along with the Office of Inspector General (OIG), published compliance guidelines for the pharmaceutical industry in the *Federal Register,* Vol. 68, No. 86, May 5, 2003. These new guidelines are called "PhRMA Code on Interactions with Health care Professionals."

Why did they need to publish guidelines you ask? You can thank the drug companies for this one. In the past, pharmaceutical companies, and other medical companies, created a whole new level of entertaining and education. Pharmaceutical companies would fly doctors and their spouses to Tahiti in order to "educate" them on why the doctor should be writing prescriptions for that company's drug. Taking twenty doctors to the Super Bowl to watch the game in a private suite was nothing. The government, and rightly so, felt like activities such as these could influence a physician or caregiver to prescribe a particular drug that they would not normally prescribe. If this drug was the greatest cure for cancer, who cares? What if it is a cheaper way to fight a disease and it is not as effective as an alternative medication, but the doctor writes the prescription anyway as a "thank you" for the trip to Tahiti or the Super Bowl? This is wrong, and this is why the government felt that guidelines needed to be in place. When I was managing the rep in a particular Midwest market, one of the clinics we called on put on a day-long lecture for people who wanted to learn more about that clinical condition. The course was on a Saturday, and in order to pay for lecture hall space, food, etc., the clinic wanted to charge all of the companies who sold to them $500 to have a "booth" at the meeting. We are not a pharmaceutical company who sets aside $200 million for marketing expenses, so we have to pick and choose which education programs we sponsor and pay to attend. This one was, in no way, shape, or form, worth the money to us. We told the clinic we would have to pass, and the rep received a message from the doctor sponsoring the meeting that "any companies that did not pay the fee, would not see their products used by this clinic." Is that pretty clear to you? I am not sure how the laws work regarding extortion, but to me this did not seem like it was much different from someone holding a gun to our head telling us to pay the $500.

This is the kind of garbage the government should be stopping, and I am guessing that PhRMA intended to help with this, but it has not yet made a huge impact. Here is another example. A major teaching hospital in a southern state decided they would create a new revenue stream, or at least that is how we looked at it. This hospital told all medical companies that in order to even walk in the door there, they would have to register as a "lobbyist," and pay an annual fee of nearly $500. You think I am joking, don't you? I wish I was. Companies with four reps/managers that needed to access the hospital would have to pay nearly $2,000 in order to sell in *that* hospital. I have no clue how this is legal, but apparently it is, and all companies had to pay the fee. Before companies paid this fee, I am sure they ran it through all legal channels to see if it was legal and allowed, and then the company wrote the check. I cannot express to you how much this bothered me, knowing that the government spends a great deal of time on health care fraud and abuse, yet allows this kind of thing to transpire. Why does it matter? I was talking to a customer about the aforementioned situation and the never-ending catering services pharmaceutical and some medical companies supply to health care. The customer responded with, "And yet we have so many people who are uninsured in this country." Hello! Is this thing on? That is EXACTLY why we have so many people uninsured. If any company has to pay to bring in lunch or exhibit at some silly education fair, the costs associated with that are added to the cost of products the hospitals or health care provider buys. This practice artificially raises the price of products and services, which raises the prices health care providers have to charge, which raises the rates insurance companies have to pay, and before you know it, insurance premiums rise to the point that some people cannot afford to pay for health care insurance.

None of this is meant to be legal advice, because I have absolutely no expertise in this area, but here are a few obvious tidbits which may help you to avoid trouble. Make sure you follow your company's policies and procedures exactly as they are given. If one of your company's policies seems wrong, talk to your superior and document that conversation. I have a close friend in the food business. He works for a manufacturer that sells specialty food items to brokers and grocery stores. I only included this tidbit in the chapter because in my friend's business, the government does not provide any reimbursement. The companies in that business have few official guidelines on entertainment, except as limited by the money they are willing to spend. Therefore, I thought it was important for you, the reader, to see how medical sales adds something else you need to think about that might have been a non-issue in your previous job.

Some medical companies do not follow the PhRMA guidelines because they feel that the guidelines only apply to pharmaceutical and biologic companies. Smart companies are implementing these guidelines anyway, because they see that the guidelines make sense and may help keep their people out of trouble. Do not give reimbursement advice. If a deal sounds a little strange, report it to your superiors. Never tell a customer that your product or service will do more than it can do. Do not ever, ever pay for business, and make sure that you are the one taking the high road and following the rules. You may lose business because you would not pay for someone's baseball tickets, but that is money well unspent. Let your competitor be the one to go before the Grand Jury. If you are ever, ever in doubt, calling the government for guidance will be of little help, just like calling the Internal Revenue Service for help on your taxes, so consult the Legal Department of your own company and follow their recommendations.

In 2005, the Advamed group (Advanced Medical Technology Association), which is a consulting group, finalized a set of rules or guidelines called, "The Code of Ethics for Interaction With Health Care Professionals." One could write an entire book on these new guidelines, but many medical device manufacturers and distributors have begun adopting the guidelines. The "Code of Ethics" includes opinions and recommendations on everything from Educational Grants to Entertainment; and from Consulting Services to Third Party Training. For a complete copy of the Code of Ethics, go to www.advamed.org/MemberPortal/About/code/codeofethics.htm. I highly recommend that you read these guidelines and conduct yourself in alliance with them. It can only help.[16]

# Chapter Eighteen

## *Now What?*

By this point in the story, you should know a lot more than you knew before you started reading this book and I am hoping that it did not take you too long to get through it. My hope is that your manager, boss, Training Department, or whomever gave you this book, did so in hopes that it could teach you the basics of what you need to know to start selling in the healthy health care market. If you paid attention in the beginning, you will remember that this was my hope as well. The question is: Where do you go from here?

If you, the reader, wanted to learn anything about a subject, you would hopefully do a little research. Let's pick a subject. Why don't we say that you saw a program on painting on one of the cable channels, and I am not referring to the painting one might do to the outside of your house; the program piqued your interest in becoming a painter, and you also think people might want to buy your paintings someday. So, what do you do? These are some things *I* would do, and not in any particular order:

- I would read a book or two on the subject so I can understand it a little bit better
- I would go out and buy some paint, brushes, and canvas or paper
- I would try to talk to some other successful painters
- I would review all my information

But most importantly, I would start to *paint*. I would paint on different surfaces, I would paint with different colors, and I would review my painting prowess after each painting. Are you getting my cute little analogy here? After you read this book, you need to start to SELL!!! Sales, like anything else, is a skill. Medical sales, is not only a skill, but it requires constant learning and review to ensure that you are on the cutting edge of information applicable to your market. How can you be an asset to your customer if you are not up on all the latest products, trends, patient issues, government/insurance guidelines, and competition? If I am your customer, and you bring nothing to the table, why should I see you? If I happen to be your boss, and you do not keep getting better and better and more and more knowledgeable, what are you doing to make me want to keep you around? If I am your husband or your wife (that might be hard for me to be your wife), where is all the great money you were supposed to be making when you went into medical sales from copier sales? Come on. Make us all proud and use this basic information to become the best medical sales rep you can be. When I sat down to write this book, as I mentioned earlier, my hope was that your boss could give you this book on Friday, and by Monday you would know all the real basics you would need to know to start making appointments and meeting with your customers, but remember that this is *only* the basics—kind of a Medical Sales for Dummies. The rest is up to you.

One other thing that is very, very important: Be nice and friendly to EVERYONE you meet in each and every health care facility you enter. You may be thinking that you will, of course, be nice to all your clinicians, but I am telling you to be nice to EVERYONE in the entire facility. Why? The following story may illustrate why this is important. If you do not get it, you need to apply for a pharmaceutical sales job.

*****

I heard a story once of a great sales guy who was working with his manager for the first time. The rep took his manager down to the lower level to check in at Purchasing in a small hospital outside of town. (Remember that "Purchasing on the lower level" stuff?) On the way to the Purchasing Office, the rep saw the maintenance guy who was painting a red stripe on the wall. This rep said, "Hey Edgar, how is the world treating you today?"

Edgar replied, "Dave, this is a good day." (The manager was thinking, how in the heck does Dave know this guy's name?)

"Edgar, how is that son of yours? Has he started high school yet?"

"Oh yeah, he's playing on the soccer team, and the kid is pretty good. Thanks for asking Dave."

"Glad to hear it Edgar. You take care of yourself."

"Same to ya Dave."

During their sales calls in the hospital, Dave and his boss saw the heads of three different departments as well as the director of Materials Management. With every sales call in that hospital, Dave asked about their families, and he seemed to show a genuine interest in what was going on in their lives. On that particular day, Dave did not actually close any new business, but he planted a lot of seeds. When Dave and his manager got in the car to leave to go to the next hospital, Dave's manager said, "Dave, I am really impressed that you have taken the time to truly get to know the director of the ICU, the Infection Control director, and even the head of Materials, but I have to ask why you wasted time getting to know the maintenance guy, you know, that guy who was painting the wall." Dave replied, "Boss, you never know, someday I might be selling paint!"

# Good Luck!
# and
# Have Fun!

# Epilogue and Opportunity

I added this section for one reason: My publisher asked me to. Despite that, this is probably something that may interest some of your friends. Perhaps sometime in the past you asked a friend or relative of yours or maybe a friend of a friend who was working in medical sales the following question, "How do I get into medical sales?" I have talked with recent college graduates with degrees in marketing, business, finance, or whatever, and some of them told me that they were either going into this business, that business, or medical sales. Or, I meet people from time to time who express to me that they have a desire to get into medical sales.

Medical sales is a great business. Very few companies with any substantial sales go out of business. Our population is aging, which means that more people are hanging around longer, which means that they need health care for a longer period of time. This is a great business. If the economy is flying high, the sale of medical supplies is high. If the economy is in the dump, medical supplies seem to sell even more. Medical sales was a good place to be in the fifties, sixties, seventies, and all the way into the current century. So, who wouldn't want to be in medical sales, if you are a salesperson? Fortunately, or unfortunately, it is not always that easy to break into this business. Due to the strong performance in this market and the solid money to be made, we find no shortage of people wanting to enter the exciting area of medical sales. If some recruiter tells you they have a magic way to move you from nothing to medical sales, they are possibly misleading you. The following is a list of some of the things I, as a medical sales manager, would look for and possibly recommend you have in your background to get a *shot* at an interview.

- Ability to learn technical concepts quickly and effectively. (How do we know this? We don't, but if you can convince me that this is a skill you possess, I may be willing to move to the next interview step.)

- Make me like you. Scratch that. Make me think that our customers will really like you. Convince me that you are personable, likeable, and that our customers would easily warm up to you.

- Even for entry-level medical sales positions, most companies want to see a track record of two to three years of successful outside sales (business to business). Selling cell phones at Best Buy or the Sprint Store does not count . . . unless you make it count. Huh? Right, emphasize any outside selling you did during your position selling cell phones, or whatever. Did you call on corporate accounts? Tell us about that. (I would suggest consulting someone who excels at writing resumes, because you will need a really good resume to get someone like me to even talk to you).

• In my division, we look primarily for people who have two to three years of successful medical sales experience. (We can afford to do this, because we pay well, and we can be choosy.)

• Try to network as much as possible. This can be an overused statement, but if I have a doctor or nurse I respect who tells me I should interview you, I will do so. The person they are recommending may not impress me on paper, but the personal relationship may get that person an interview. If you are the person getting that interview, you have to make it count.

• If you are not in medical sales, and you have read this book, I would ask your interviewer why they want you to have experience in medical sales. I would then tell them that even though you lack experience in medical sales, you have a good basic understanding of medical sales, because you were smart enough to read this great book!

If you are lucky enough to get the job, remember that it is up to you to make this tremendous opportunity count. Do not disappoint me. I gave you the basics of what you need to know, so put it to good use.

# Bibliography

1. NCRPM Report, "Structural Shielding Design for Medical X-Ray Imaging Facilities," No. 147 (Bethesda, M.D.: NCRP,2004).

2. *Journal of the American Dietetic Association,* 105:1, pp. 139–144.

3. "Our History," Cardinal Health website, www.cardinal.com, 2007.

4. Owens & Minor, *2006 Annual Report,* p. 2.

5. "About Us," McKesson Empowering Health care website, www.mckesson.com, 2007.

6. "Fact Sheet," Novation website, www.novationco.com, 2007.

7. "About Ascension Health," Ascension Health Website, www.ascensionhealth.org, 2007.

8. National Account Information, National Accounts, General Questions, Medline Industries, pp. 4–11, 2006.

9. OIG Compliance Program Guidance for Pharmaceutical Manu-facturers, Federal Register, Vol. 68, No. 86, Monday, May 5, 2003.

10. *Webster's New World Dictionary,* ed. 4 (New York: Wiley, 1999).

11. "Medicare and Social Security (Full Retirement) Eligibility," Centers for Medicare & Medicaid Services,

CMS Pub. No. 11038, Revised July 2006.

12. "The Home Health Pay for Performance Demonstration," Demonstration Overview, Centers for Medicare & Medicaid Services, October 30, 2007.

13. HCPCS Coding example, CMS.gov.

14. "Trends in Indian Health," National Library of Medicine, 1998–99.

15. "Recognizing Hospitals with the Highest-Quality Care and Greatest Operational Efficiency in the Country," healthleadersmedia.com, July 7, 2007.

16. Advamed Code of Ethics, Advanced Medical Technology Association, Washington D.C., 20004, www.advamed.org.

If you need to reference the book, Van Vechten, Lee R., *The Successful Manager's Guide to Selling Through Proactive Customer Service* (Omaha: Business By Phone, Inc., 2002).

# Index

# About the Author

Michael Carroll graduated from Lindenwood University with a degree in Business Administration and lives in the Midwest with his wife and daughter. Carroll has been selling in the health care field for over twenty years and currently is a regional manager for a major medical company. Michael has trained over 100 medical sales reps and currently teaches the product sales training class for his company. In his spare time, Michael enjoys sports, spending time with his family, playing drums and writing.